Reducing Operational Costs in Composites Manufacturing

Reducing Operational Costs in Composites Manufacturing

Karen Snyder Travis

CRC Press
Taylor & Francis Group
Boca Raton London New York

CRC Press is an imprint of the
Taylor & Francis Group, an **informa** business

CRC Press
Taylor & Francis Group
6000 Broken Sound Parkway NW, Suite 300
Boca Raton, FL 33487-2742

© 2020 by Taylor & Francis Group, LLC

CRC Press is an imprint of Taylor & Francis Group, an Informa business
No claim to original U.S. Government works

Printed on acid-free paper

International Standard Book Number-13 978-1-138-60888-7 (Paperback)

Library of Congress Cataloging-in-Publication Data

Names: Travis, Karen Snyder, author.
Title: Reducing operational costs in composites manufacturing / by Karen Snyder Travis.
Description: Boca Raton, FL : CRC Press/Taylor & Francis Group, 2019. |
Includes bibliographical references and index.
Identifiers: LCCN 2019009995| ISBN 9781138608887 (pbk. : acid-free paper) |
ISBN 9781138609051 (hardback : acid-free paper) | ISBN 9780429466397 (ebook)
Subjects: LCSH: Composite construction–Cost control. | Manufacturing processes–Cost control. | Composite materials industry.
Classification: LCC TA664 .T73 2019 | DDC 620.1/18068–dc23
LC record available at https://lccn.loc.gov/2019009995

Visit the Taylor & Francis Web site at
www.taylorandfrancis.com

and the CRC Press Web site at
www.crcpress.com

For Dave

who gave me the inspiration to write this

book

Contents

Preface ix
Acknowledgments xi
Author Biography...................................... xiii

Section 1   Understanding Your Weaknesses 1

1 The Basics of Composite Manufacturing........................3
2 Document the Baseline 11
3 Evaluate the Data 15
4 People, Processes, and Materials 23

Section 2   Getting Organized 37

5 Know Where You Want to Be................................39
6 Developing the Plan to Achieve Your Goals 45
7 Clean It Up... 51
8 Work the Plan 55

Section 3   Hiring and Keeping the Best Employees 59

9 Character Skills vs. Experience 61
10 Motivation 67
11 Choosing and Empowering Leaders 75
12 Support the Team 83

Section 4   Utilizing Free Support......................89

13 Inside Influence 91
14 Co-Managers 97
15 Outside Influence 103

Section 5   Sustain 127

16 Measure 129
17 Accountability... 139
18 Reward 147
19 Change Management 157

Section 6   Forward-Thinking 163

20 Preventing Major Issues 165

Glossary 177
References 187
Index 189

Preface

Over the course of the past 30 years, I have had the privilege of setting up or managing departments and plants in the composite industry and/or helping other managers solve problems and reduce operation cost in their facilities. As a manager, I began most positions with a feeling of both anticipation and trepidation. I can honestly say that I have made my share of mistakes and have witnessed the mistakes of others. It was through these mistakes and often successes, I have learned to fine-tune what works, and avoid what does not.

The book is divided into six sections and is written with organization-specific principles utilizing a "how to" format for new or struggling managers in composites. There are several examples given in each section, which support the points stated.

In the first two sections, readers learn how to evaluate the existing environment for determining the best course of action when developing a plan to achieve goals. This is followed by a deeper understanding of how to effectively manage employees and why character strengths are important in section three. Section 4 helps the new manager to think outside the box by bringing in other managers to evaluate and offer suggestions. Section 5 teaches the reader how to sustain and continually enhance what they have put in place, and finally, Section 6 outlines the importance of utilizing the support from two departments that often get overlooked as avenues for reducing operational costs: Maintenance and Quality Assurance. This section helps to set up these departments for the necessary help to attain future success in manufacturing.

It is important to understand the difference between knowledge and wisdom. You can achieve all of the organizational skills required to perform the responsibilities of a manager, but if you do not acquire the wisdom to use these skills and to deal effectively with the many different employee personalities, your success will be greatly limited, and short-lived.

A tool is a tool, its usefulness is defined; but the human capacity has yet to be limited. Simply put "People can make you or break you". When your goals are centered around helping others be successful, you stand a much better chance at being successful as well.

I would like to thank the following experts for contribution of examples throughout the book:

David L. Travis
General Manager, Composite Research, Inc.
Blackshear, GA

Tim Reid
Process Control Manager, Monterey Boats
Williston, FL

Henderson Clarke
Lamination Supervisor, Marine Concepts
Sarasota, FL

Andrew Pokelwaldt
Director Certification, ACMA
Arlington, VA

Julie Jenewein
Sourcing Manager, American Bath Group
Savannah, TN

Richard Ryan
President, AIM Supply
Largo, FL

Mark P. Guilfoyle
Territory Manager, Norton
Saint Augustine, FL

Marshall "Ty" Ellis
VP Business Development, Vallen
Jacksonville, FL

Alain Lacasse
Owner, Nano Boats
Venice, FL

Acknowledgments

I would also like to give a heartfelt thanks to two special people that although they did not directly contribute to this book, they definitely contributed to my ability to write it. I consider both of them mentors to me as I have been blessed to learn so much from both. Don, whom I can never speak with again, and really didn't get the chance to properly thank him for teaching me how to think like the senior management team; and Joe, who still mentors me, although I'm pretty sure in his humility, he might not agree.

Donald D. Wakeman, Naval Engineer, Plant Manager, Composite business Owner, Consultant

Joseph J. Mikolajczyk, Electrical Engineer, Manager Segment Marketing – Energy, Marine

Author Biography

Karen S. Travis has earned an Executive MBA, and a Bachelor of Arts degree in psychology, from the University of South Florida. She has earned certifications in ABYC, NMMA, CCT, ISO 9001, and is a licensed Realtor and published author.

Mrs. Travis has worked in and managed several areas in the composite marine manufacturing industry including Quality Assurance, Production Control, Upholstery, Lamination, and as a Plant Manager. She previously worked with a composite consultant on a two-year project for starting a boat manufacturing facility in Russia, and has compiled and written seven manuals to support this.

In the past 13 years, she has worked for Saint-Gobain Corporation and currently holds a position as an Account Manager for the abrasive segment (Norton, Carborundum, Merit) calling on the marine and composite customers, and works closely with them to solve problems, and reduce operational costs.

She has spoken on several occasions as a presenter at industry trade shows including IBEX and CAMX concerning Best Practices for using Abrasives in the Repair of Gelcoat, and Reducing Operational Costs in Composite Manufacturing, and is currently working on a seminar topic titled: Implementing an Effective Quality Assurance Department for Composites.

Section 1

Understanding Your Weaknesses

1

The Basics of Composite Manufacturing

Introduction

At a minimum, a general understanding of the various departments in composite manufacturing is helpful or necessary for developing the most advantageous plan to reduce operational costs. The following departments are explained briefly in this chapter and will be addressed in more detail in other chapters throughout the book. They are divided into three groups: departments that support the manufacturing processes including safety, maintenance, engineering, procurement, production control, quality control and quality assurance, and warrantee; departments that manufacture the product including mill, tooling, gelcoat, lamination, pull and release, trim and grind, and repair; and departments that add to or complete the manufacturing processes including but not limited to sub-assembly, assembly, upholstery, and shipping.

Support Departments

These departments do not add labor or materials to the completed product but do affect the ability to perform the manufacturing and production processes in a controlled, repetitive, and consistent way.

- Safety
  Safety is one of the most critical departments in a composite facility and can be the costliest if left unchecked. The Safety department is responsible for working with the plant management to make sure the employees understand the hazards they work with and adhere to the safety practices and government laws. They are also responsible for working with the materials and maintenance groups to be certain the safe handling of all chemicals is followed. The costs from not following safe work practices can range from minor fines to complete loss of a composite facility, or death of an employee.

- Maintenance
  The Maintenance department is key to keeping the facilities in operational order. This should include all building structural, electrical, mechanical & plumbing, as well as all equipment, machinery & tools. Preventative maintenance is essential to reduce downtime events. The Maintenance manager should track all spending in this department and should report monthly to the plant manager regarding any issues from the workforce on the care of equipment, machinery, tools, and the facility.

- Engineering
  The Engineering department develops new product and adjusts as necessary, which includes all molds, jigs, fixtures, guides, and patterns. This department is also responsible for making sure all production parts are assigned a part number, a Bill of Materials, and the allocation of labor hours to manufacture.

- Procurement
  The Procurement department is responsible for ordering, receiving, storing, packaging, and delivering all products used to manufacture the end product. They take information from engineering, and source at least two vendors for each product. Their goals are to keep the cost of the products at a level commensurate with the goals set by engineering and keep the stock at a level that allows for Just-In-Time (JIT) deliveries without production interruptions.

- Production Control
  If you think of the plant in the likeness of a person, the Production Control department would certainly be the brain. This department is key to keeping everything in the plant flowing without hiccup or bottleneck. Production Control defines labor at work centers, schedules all work in a balanced timeframe, and determines completion dates on all manufactured product. They work closely with Engineering, plant management, sales, and procurement to assure a smooth production flow, and provide the necessary paperwork and schedules for each area of the plant.

- Quality Control/Assurance
  Quality Assurance is the responsibility of every person working in the facility. Each individual that touches any part of an end product or works in an area that supports the manufacturing process, is responsible for making sure everything they do for the company is 100% accurate.
  Quality Assurance can be performed formally with checkpoints and sign offs at every function, or it can be an honor system with effective leadership checking sporadically during the process to be sure the product is manufactured according to the quality standards.
  Quality Control is performed after the process is completed and is

usually done by an inspector working in a Quality Control department that usually does not answer to the plant manager. Although product quality is the responsibility of the plant manager, it can be a conflict of interest to have this department working directly for the person responsible for the production of finished product, especially if the owners or officers of the company places more value on completions dates than quality.

- Warrantee
  Although this department is not directly involved with the manufacturing of composite product, often the replacement or repair of sold product is given to the manufacturing process to complete, and this can interrupt the normal process flow of manufacturing especially where tooling and mold availability are not commensurate with product manufacturing requirements. An example of this issue happens when a warranted composite component part must be added to the production schedule, but the production schedule for that mold does not allow time for additional product.

Manufacturing Departments

These departments are directly involved in the manufacturing process and add labor and materials to the completed product.

- Mill
  Often called the "Wood Shop" or "Fab Shop", this department is composed of equipment and tools that cut and process raw core materials such as wood, composite sheets, plastics, fiberglass, and other structural materials used as support structures applied in the lamination processes of certain types of composites. A good example is the stringer system which is laminated into the hull of a boat.
  This department may also manufacture support structures for other departments, which add to the manufacturing processes such as core material for upholstery or completed cabinetry for assembly functions. As core and support materials have changed from using mainly wood to including other composite materials, the name of this department also changed from wood shop to fab shop or Mill.

- Tooling
  The Tooling or mold department is responsible for maintaining and preparing the molds for production. The supervisor of this department should keep a log file on all molds, which identifies how the mold was built, any repairs performed on the mold, and the wax cycle.

- Gelcoat
  The Gelcoat spray department takes the prepared molds from the tooling area and prepares them for the gelcoat spray. The department is responsible for maintaining the spray equipment and calibrate the guns at every shift. They are also responsible for checking to be certain all controls are in place prior to spraying, which includes visual inspections of the equipment, spray area, chemicals, and molds.

- Lamination
  Lamination applies layers of materials that bond together to give the end product strength and stability. This can be done in several ways such as filament winding, spin molding, open molding, closed press molding or squash molding, infusion, injection, resin transfer moulding (RTM), and several other ways. The Lamination department is responsible for working with maintenance and should calibrate all equipment as required at every shift. They are also responsible for checking to be certain all controls are in place prior to the lamination process, which includes visual inspections of the equipment and chemicals used.

- Pull/Release
  This department releases the part from the mold. They are a key factor to the Tooling department in determining if there is an issue with the release agent used. If a part has a hard time releasing from the mold, it may indicate that it is time to spot treat or strip and re-wax the mold. They should indicate this in the mold log file. See Figure 1.1 for an example of what this looks like.

- Trim/Grind
  Trim & Grind removes the excess laminate flanges and flashing from the part. In complex manufacturing processes, they also cut out all access holes and grind all rough edges. This is done in a controlled area as the dust can be combustible and care must be taken to prevent fire.

- Repair
  The Repair department fixes all gelcoat defects resulting from the previous production processes.

Production Departments

In a complex composite manufacturing process such as boat, automotive, aerospace, or wind energy manufacturing, the completed product may include the assembly or installation of hardware, cabinetry, electronics, or other items which require additional processes to complete the finished product.

Mold # D-21SP-02
Master/Plug # M-21SP-01
Manager John Doe

| | | | Time | | | | | | Operation Specifics | | | | | | | | | QA Controls | | | |
Date	Ambient Temp (F)	Humidity	Start	Stop	Stage	Mold temp	Chemical	Material	Catalyst type	Catalyst level	Mill/oz Wet	Gel Time	Peak Exotherm	Barcol-1HR	Gloss	QA	Equip Calibration	Mtl Handling	Application	Visual
9/10/12	76.0	89.0	8:02	8:32	1-Gel	76	Polycryl G501	N/A	Hi-Point 90	2.0	42	22	300	42	N/A	Jane Doe				
9/10/12	92.0	89.0	10:30	11:05	2-Skin	120	Ashland vinyl ester 601-200	Chop.												
9/10/12	94.0	88.0	12:50	13:20	3-Skin	120														

FIGURE 1.1
Mold Log File

- Sub-Assembly
  This department assembles smaller parts, which then are installed into the completed product. A dash panel, console, or seat assembly are good examples.
- Upholstery
  Upholstery takes raw materials such as fabric, foam, and core, and cuts, sews, and assembles these into upholstery parts that are given to the final assembly areas to complete the finished product.
- Assembly
  The Assembly department takes all manufactured and assembled product and completes the finished product.
- Shipping
  The shipping department takes the finished product and prepares it for shipping to the customer, which can include assembling the necessary paperwork for the customer, the transport, and the government.

Conclusion

Understanding what the different types of departments are and how they operate in a composite account will help you develop a stronger plan for reducing operational costs in your plant. Some of the departments are specific to the industry, while others can be found in most manufacturing companies, but they all must work together, as in any plant, if you will have the greatest success for reducing costs.

Terms

calibrate
closed press molding
composite
filament winding
gelcoat
infusion
injection
lamination
open mold
procurement
RTM
spin molding

squash molding
stringer system
tooling

References

www.google.com/search?safe=strict&client=safari&site=async/dictw&q=Dictionary
www.merriam-webster.com/dictionary/term

Additional Reading

www.advanced-plastics.com/fiberglass/Composites_Application_Guide.pdf

2

Document the Baseline

Introduction

Generally speaking, if you have found yourself in this position, it is more than likely because you have shown that you possess at least some of the skills and education required to manage a facility, as well as organize and lead a team, although I have worked in places where this was definitely not the case concerning some of the promotions.

Whether you are new to the position of manager or have been in the position for years, understanding in detail where you are is imperative to developing a plan to reduce operational costs in your facility. In this chapter we will discuss how you need to assess the existing situation of the plant, and which reports and documentation you will need to help you evaluate how the plant has been performing.

Get to Know Your Plant

Start by spending a week just observing and learning. Walk around the plant and utilizing your phone's camera, note everything and anything you see which does not make sense to you. No matter what, write it down and if applicable, photograph it. Make sure you do not miss any area. Treat this plant as if you just purchased it and all it contains. Some day in the future it will be a great show and tell, or a before and after story, and it will also help you to stay focused when you are having trouble dealing with the issues. Documenting the existing situation is imperative to understanding the order and assignment of resources for change, and for getting your employees engaged in the change process.

Be sure to not miss the opportunity to greet as many of the employees you meet as you are able, and speak to them. Get to know them, what their job function is and how they like it? Careful do not take notes here, you are only getting to know the people in your charge and they you. You may or may not remember their name until you have greeted them a few times, don't worry about this, it is just important that they get a feel for who you are. Try to have a conversation with a few of them. Ask questions. You don't know everything; you don't really know anything yet about this facility, so you can learn a great deal from the people in your charge. They will feel valued, and will respect you for it. Make sure you are sincere. If you greet them again and cannot remember their name, simply ask them apologetically again. There is a great deal to be gained from admitting our shortcomings, you will be teaching your employees that it is indeed human to error, and honorable to own up to it. Guess what? You are not perfect, and they aren't either. Learn to laugh at yourself occasionally.

The pictures you take will serve as a starting point, a beginning; something for you to measure against while tracking cleanliness, safety, and organization. You will want to save these to a file, and if appropriate, post them to a wall in your office. Be careful not to give negative comments about the condition of the facility to your employees or comrades. Your employees may have been fond of the previous management, and will take any negative comments regarding it as an insult. This will not help you.

You will need to take this walk almost every day for the remainder of your career. After the first few times you will not need to take pictures, but it isn't a bad idea to occasionally photograph the workers from time to time, especially if producing something new, or interesting to talk about. People like to be recognized, this will build morale. You will also notice things, which you have missed in the past, or which have recently cropped up. It is also a great idea to occasionally put pictures up in the break areas, and let the employees derive their own opinions.

Gather the Existing Measurement Reports

Next, and unless you want to reinvent the wheel, you will need to understand how the plant has been functioning. So, start by gathering the performance information from the previous 6 months. Below is a list of the key reports you should be familiar with.

- Performance Efficiencies
- Material Usage & Waste
- Shop Supply Usage & Waste
- Bill of Material Issues
- Safety Incidence & Training
- Product & Performance Quality
- Warrantee
- Area Budgets
- Labor & Overtime Usage and Variance

- Employee Retention & Absenteeism
- Customer Satisfaction

Most companies keep this information electronically, which will make this task pretty easy. Just get the IS department to supply the previous bullets. Or if not, have the various department heads prepare this information for you. Regardless of how you get this information, it will be needed in order to have an in-depth understanding of how the plant has been performing in order to develop the most successful plan to improve it.

Conclusion

Whether you are new to the position of manager or have been in the position for years, understanding in detail where you are is imperative to developing a plan to reduce operational costs in your facility. In this chapter we learned how to assess the existing condition of your plant, and which reports are necessary to understand how the plant has been performing. This information will be used to evaluate the plant's performance in more detail to gain a better understanding of how to put a plan together to reduce operational costs.

Terms

Operational Costs
Morale
Performance Efficiencies
Employee Retention
IS department

3

Evaluate the Data

Introduction

An in-depth understanding of the plants performance is derived from the notes and pictures you took, and from evaluating the performance data from the reports you have gathered from each department, including Performance Efficiencies, Material Usage & Waste, Shop Supply Usage & Waste, Bill of Material Issues, Safety Incidence & Training, Product & Performance Quality, Warrantee, Area Budgets, Labor & Overtime Usage and Variance, Employee Retention & Absenteeism, and Customer Satisfaction. These reports are described in detail below according to their function and usefulness for helping to understand how the plant has been performing. This information is critical for developing the most successful plan for reducing operational costs in your plant.

Evaluating the Reports

Once you have gathered the necessary information, you will want to put it into a database program such as Excel for creating trend-line graphs. An example of a Performance Efficiency trend-line graph is depicted in Figure 3.1.

You will be adding to these graphs as you track your reporting to the above bullets each week.

An in-depth understanding of each of the reports is necessary for recognizing the impact this data has on the bottom line, and to develop the most beneficial plan to reduce costs in the facility.

Efficiency Reports

Generally, the Engineering department in a facility will produce a product to be manufactured, which will include Build Books, Bills of Materials, Standard Hours for each job function, all fixtures and jigs necessary to perform the task, and cost based off of labor and materials necessary to

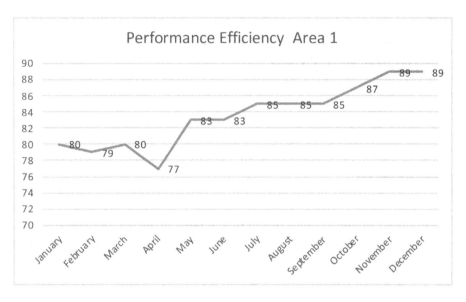

FIGURE 3.1
Performance Efficiency Trend-Line Graph

perform the task. The Standard Labor Hour is used as a base to compare against actual labor hours for tracking efficiencies, so you and your employees can understand their performance and how it relates to improvements. See Figure 3.2 for an example of an efficiency chart by area, by month.

Department: Lamination	Std Hrs	Hrs used	Std Hrs	Hrs used	Std Hrs	Hrs used	Std Hrs	Hrs used
Operation	Jan	Jan	Feb	Feb	Mar	Mar	Apr	Apr
Gel Spray	1125	1172	1035	1163	1238	1360	1114	1198
Skin	6250	6510	5750	6461	6875	7555	6188	6653
Bulk Lam	11250	11719	10350	11629	12375	13599	11138	11976
Stringers	7962	8294	7325	8230	8758	9624	7882	8476
Pull	1013	1055	932	1047	1114	1225	1003	1078
Trim	2111	2199	1942	2182	2322	2552	2090	2247
Repair	1980	2063	1822	2047	2178	2393	1960	2108
Total	31691	33011	29156	32759	34860	38308	31374	33736

FIGURE 3.2
Efficiency Chart

I have tracked the efficiencies of several department in many facilities and have heard many excuses such as "the accounting method is inaccurate, so the efficiency number is also inaccurate", or "the standard hours are inaccu-rate", and "we are missing parts in the Bills of Materials that do not include standard hours, and this goes against our efficiencies". All of these statements were true, and the efficiencies were not accurate; however, what I want to stress here is that the current efficiency number is only a starting point. If you do not change the accounting method, and I suggest you do not for at least the first 3 months, then it really does not matter if the accounting is inaccurate as long as you show a constant improvement. Don't hold the department to a 93.75% efficiency (7.5 hours available/8 working hours) until you know the data is as accurate as possible.

I managed an area in upholstery a few years back where the Bills of materials were highly inaccurate, to the tune of 50%. We produced many parts, which did not exist in the system. Not only were the efficiencies skewed due to this, the material accountability was as well. Once we had the employees under control, and the production rate under control, we then started to work on the details of adding parts to relieve materials and labor. We took this plant from 80 employees making 18 boats daily at 49% efficiency with a material variance of around 40%, to 97% efficiency with 48 employees building the same amount of boats, and a material variance of around 5%.

Material Usage Reports

This report will show you how much material you are using in comparison to how much is required to build the production schedule. If the Bills of Material are accurate, the deficiency is called waste. It's your job to identify and measure waste and find ways to eliminate it as much as possible. See Figure 3.3 for an example of how this report is used.

Material Usage Jan 2019	BOM LBS	Actual LBS	Variance LBS
Fiberglass	10325.00	10118.50	206.50
Resin	6711.25	6845.48	-134.23
Gelcoat	805.35	808.57	-3.22
MEKP	134.23	136.91	-2.68
Foam	1032.50	1022.18	10.33
Core	5162.50	5059.25	103.25
Total			179.94

FIGURE 3.3
Material Usage Report

Shop Supplies

Shop supplies are generally items used to manufacture the products produced, but generally are not included in the product when complete such as cleaning agents. On occasion, I have seen composite facilities include sealants into this category as it is difficult to determine how much per part is needed. Once this is determined, the actual usage should be compared to the base (see Figure 3.4).

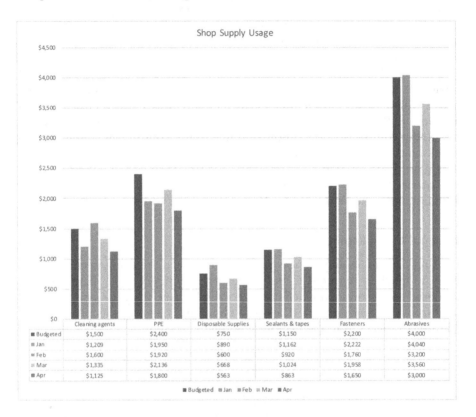

FIGURE 3.4
Shop Supply Usage Graph

I often asked my new hires what they believed we manufactured and was always given the response "boats". At this point, I usually held the picture of the owner of the company, and introduced him/her to the employees, always followed by a short lecture of what we actually manufactured for the owner … . money.

To this I generally received strange looks. People usually do not go into business to make anything unless they are sure they will be making money. The end goal is always the same … . money. The sooner you are able to help your employees understand this basic fact, the easier it will be to get them to help you manage out waste. Waste, which equates to lost money, is easily found in many forms in the procurement and management of materials. But to start, let's look only at the initial coast of the shop supplies, how they are used, and how we initially feel they should be used.

Example: You have three gelcoaters, five chopper operators, twenty-five laminators, seven finishers, five grinders, and fourteen assembly workers in a small plant. What type and how much of each PPE (personal protective equipment) will be needed by these employees in a month's time-frame? This is an answer, which you should attain, and help your employees understand and attain it as well. If your budget is $6,000 per month for PPE (personal protective equipment), it should be based off of how many employees you have and what type of PPE is needed for them to perform their job. They should be allotted these supplies, and spending should be tracked. See Figure 3.5 for an example of a supply list by area.

Supply Requirements	Position Title	Gelcoater	Chopper operator	Laminator	Puller	Grinders	Finisher	Assembler
Supply Type - PPE	Quantity of Position ---- Total Supply Required	3	5	25	1	5	7	14
Respirators	11	3	3			5		
Face masks	10			3			7	
Ear Plugs	20				1	5		14
Paper Suits	11	3	3			5		
Paper Booties	14	3	3	3		5		
Spray Socks	8	3				5		
Gloves-plastic	21	3	3	3		5	7	
Gloves-leather	1				1			
Safety Glasses-Glass	9	3	3	3				
Safety Glasses-Plastic	27				1	5	7	14
Face Shield	5					5		
Aprons	10			3			7	
Paper Shirts	3			3				
Hairnets	35		3	25			7	
Safety Helmets	4				1			3

FIGURE 3.5
Shop Supply List

BOM Issues

This figure will be based off of how many boats are produced. How is your stock handled? Just-in-time? Kanban? Issued to an area/department, or to the hull number. How are damaged goods tracked? How are raw chemicals received and usage tracked?

Safety

How many lost time accidents have occurred in the last week? Month? Year? Is it posted where the employees and everyone who walks into the plant can see and understand it? What was the actual cost to the facility? Are employees rewarded for being safe? Are safety discussions and meetings held regularly? Are employees acting as safety representatives in each area of the plant? Are inspections of all safety equipment performed regularly? How often and by whom? At a very minimum, the "Days without a lost time accident" sign should be posted where all employees can see as they enter the facility.

Quality/Warrantee

Are formal inspections performed on the work? Are documents kept? Are defects tracked by product/area/employee? Are rewards given for exceptional quality work? Is feedback generated from customers filtered back to the employees in the plant, especially the ones who originated the defect?

Budget Figures

Is a formal budget defined for your plant, and is this used as a base for the next year? Are you the person who is responsible for your budget, or is it given to you to determine?

Labor

Are there labor charts available to determine how many employees are needed for each function according to how much product is produced? If there are no labor charts, you will need to create one almost immediately. See Figure 3.6 for an example of a how to determine workload labor needed.

Department	STd Hrs = 1 unit	Daily Hrs Available	Job Description	Employees Needed
Material Cut	2.00	20.00	Cutter	2.7
Gel Spray	0.75	7.50	Gelcoaters	1.0
Lamination	14.25	142.50	Laminators	19.0
Stringers & Floors	8.00	80.00	Installers	10.7
Pull	1.00	10.00	Pullers	1.3
Trim & Grind	2.00	20.00	Grinders	2.7
Repair	2.00	20.00	Patchers	2.7
Total	30.00	300.00		40.0

Production Rate - Daily	10.0

FIGURE 3.6
Assigned Labor

Employee Turnaround and Absenteeism

Human Resources should be able to provide reports per person and department on employee turnaround and absenteeism. Happy employees are generally at work on time in their jobs. If not, you will want to understand why, and put a plan together to change. Getting people to

Issue	Area Owner	5-Why Questions	Urgency (1-10)	Ease of Correction (1-10)	Cost to Change (1-10)	Cost of Opportunity Lost (1-10)	Total Score
Excessive WIP - 60K	Ronnie	1. Why is there excessive WIP? To supply the main plant. 2. Why does the main plant need excessive parts? Damage or lost. 3. Why are the parts damaged or lost? No accountability. 4. Why is there no accountability? No formal scheduling system in small parts. 5 Why is there no formal scheduling system? None has been created.	10	3	1	8	22
Materials are not delivered to a holding station	Sally		1	2	3	4	10
Excessive chop and fiberglass waste on the floor during lay up process	Ronnie		3	2	1	7	13
Several cutting jigs and fixtures do not exist or need replacement	Ernie		5	4	3	2	14
Ceiling lights are broken in Lay-up	Marty		7	4	3	2	16
Production scheduling system does not exist	Betsy		8	7	5	4	24
Shop Supplies are handed out when requested	Sally		6	9	5	9	29
Formal QA chemical testing does not exist	Stan		5	4	3	2	14
No calibration log books for chop guns	Marty		7	4	3	2	16
Quality defects average 10 per unit	Stan		8	7	5	4	24
Materials are not delivered at appropriate times	Sally		6	9	5	9	29
Shop supply cabinets are overstocked	Sally		3	2	1	7	13

Scale: 1-10, 10 = Highest weighted value
Ronnie - Layup supervisor
Marty - Maintenance Supervisor
Stan - Quality
Betsy - Production Control
Sally - Materials

FIGURE 3.7
Issue Matrix

want to change is accomplished through understanding their concerns and giving them a belief that you care and can make a change yourself.

Customer Satisfaction

If what we manufacture is money, then our survival is through our customers: those people who receive the products we make. You should be able to get quality and customer satisfaction reports from the top-level Quality Assurance manager. But let's not forget the in-line customer, those departments within our organization which we supply. You will find their opinions of your services invaluable for tracking progress.

You will need to create a spreadsheet for inputting the issues you uncovered from your walk through, and you need to categorize these issues as well by area, urgency to correct, ease of correction, and cost to change or cost of opportunity lost. This will help you determine how much and which resources you will need to correct the issues. See Figure 3.7 for an example.

Conclusion

You should now be familiar with the types of reports that identify how your plant has been performing in each area, and how the issues you have uncovered will be used as solutions for improving on the performance in these areas. This information will be touched on again in Chapter 5, and incorporated into a plan in Chapter 6, but first we need to assess the team in more detail in order to be certain we have the best team for the greatest chance of success.

Terms

Bills of Materials
Build Books
Chopper operators
Employee turnaround
Finishers
Gel-coaters
Grinders
Just-in-time
Kanban
Laminators
Material variance
Production rate
Standard Labor Hours
Trend-line graphs

4

People, Processes, and Materials

Introduction

Understanding the responsibilities of the people who work for you is key to holding them to a standard. Often leaders are given a position, and told to lead a team of individuals, but aren't given clear objectives, or don't possess the necessary qualities to be a leader. This chapter identifies how to determine if your team is equipped to perform their functions, and if they are the team to lead change for reducing operational costs.

This chapter also discusses how to determine if the existing processes for manufacturing are accurate and functional, and if there are other manufacturing processes needed in the facilities to help reduce operational costs.

The handling of materials including staging and deliveries is critical to composite manufacturing. Inefficient or ineffective handling can cause manufacturing bottlenecks and warrantee issues later. Having the correct materials in the appropriate space at the desired time is key to at least keeping operational costs in line with the standard.

Understand the Responsibilities of the Leaders Who Work for You

The people who work for you can make you or break you. We will discuss appropriate personalities per positions in Section 3, but first you will need to learn who these people are, and what is expected of them in their current positions.

Start by observing what they do, and don't forget to ask questions of them regarding their work. Just because you are their boss does not mean you have more knowledge than they do. You have a great deal to learn in your position, and your people hold many of the answers.

Take very detailed notes, and practice reserve regarding immediate solutions to every issue they give you. Give yourself time to investigate every angle to develop the best solutions for the issue. Often this takes additional information from others in the leadership team in place. Your notes will be helpful later as you read over your thoughts and comments when you are alone and have time to reflect.

You will also want to document the positions and responsibilities into a spreadsheet such as the one pictured in Figure 4.1.

Later you will use this information to help set up your organizational chart and choose the appropriate individuals per position. You may find they are best suited in the positions they hold as well.

Dig into the Procedures and Processes

How things get accomplished from company to company is a direct reflection of the procedures and processes the company upholds.

Start in the engineering department. Do Build Books or Construction Manuals exist for the parts produced? How accurate are they?

- Do they have complete materials lists with part numbers?
- Do they have dimensions?
- Are they current (within the last model year)?

Test a few complex parts with the build books or construction manuals associated to it. I have seen many cases where what the engineering department has outlined in the Build Book is not what actually happens on the shop floor. If the information provided by engineering is not accurate, chances are the Bills of Materials are also not accurate, and your efficiencies and material usage reports will reflect a variance.

You can also determine if procedures are being followed, by watching how your employees complete their tasks. Do they follow the same format when building the same part? Is the work completed the same from employee to employee? If not, you either do not have procedures, or your employees are not held accountable to them. It will be nearly impossible to expect an acceptable efficiency unless you determine the procedures and hold the employees accountable to them.

If you cannot get the support from the engineering department to set down the procedures or create the processes, then you will need to hire someone with these talents to do this work for you. Read Example 4:1 to understand how this helped reduce cost for a large manufacturer of boats.

Example 4:1   Small Part Manufacturing

The Problem: In a large manufacturer of fiberglass reinforced plastic (FRP) small parts, the engineering department had a habit of assigning the same Bill of Materials to like parts. For instance, most of the nonstructural hatches or FRP storage lids required the same or similar

Position	Name	Responsibility
Supervisor - Lamination	Ronnie Haas	Manage the direct reports to be certain the schedule is attained
		Make sure the employees follow all safety standards, and process materials to specification
		Make sure the chemicals are stored, used, and waste disposed of properly to prevent fire
		Handle all employee issues in your charge.
		Deliver quality product to the areas you support
		Follow proper start up and shut down procedures.
Lead- Molds & Gelcoat	Gilligan Anderson	Prepare all molds for gelcoat spray.
		Strip and rewax molds as required
		Repair all damaged molds and report damage to the supervision
		Follow proper start up and shut down procedures.
Lead- Lay Up - Consoles & Hardtops	Ginger Rodgers	
Lead- Lay Up - Stringers & Buckets	Maryanne Knight	
Lead- Lay Up - Seat Bases & Hatches	Jose Sanchez	
Lead- Foam & Pull	Steve Martino	
Lead- Trim & Grind	Skip Tripp	
Supervisor - Sub Assembly & Repair	Fred Stone	
Lead - Patch: All FRP Parts	Larry Peters	
Lead- Console Assembly	Wilma Redman	
Lead- Hatch Assembly	Peggy Unger	
Lead- Hardtop Assembly & Shipping	Rosita Vasquez	
Lead- Seat Base Assembly	Barney Short	
Lead - Shipping	Alexis Roberts	
Plant Manager	TBD	
Lead - Maintenance & Safety	Robert Diamond	
Lead - Quality	Penny Wise	

FIGURE 4.1
Employee Descriptions.

materials, and were relatively similar in size, so instead of documenting the exact materials and time to build, the Bills of Materials were copied to save time. This created a material and labor variance, and also made it challenging to accurately price replacement parts for customer service and warrantee. See Figure 4.2 for an example of what a Bill of Materials looks like for a specific part.

Completed Product Number   111-222

Part Description   Hatch

Labor   49 Minutes

Part #	Description	UOM	Qty
230-678	Gelcoat	LB	2.5
345-678	Chop Strand	LB	4.2
456-789	Resin	LB	9.8
342-435	MEKP Clear	LB	0.04
342-567	MEKP Red	LB	0.13
789-098	Hinges	EA	2
123-234	Latch	EA	1

FIGURE 4.2
Bill of Materials.

The Solution: When this issue was brought up to the engineering team, it was not addressed as resources were needed in other areas that were considered more important. It was obvious that correcting the issue would be the responsibility of the plant.

As with any issue of this nature, you need documentation to support what is expected, documentation to support what is actual, and a very good understanding of what resources you have to support the variance.

Engineering provided us with the documentation to support what was expected, we determined that we would need to weigh and measure all materials and completed parts for five of each specific hatch built, and provide detailed time studies for the same. We would also need to determine which individuals in the plant would perform this work without creating a deficiency in the plant production, and decided the fiberglass kit cutting operation would measure and weigh the raw glass materials; with the help of attached scales (see Figure 4.3

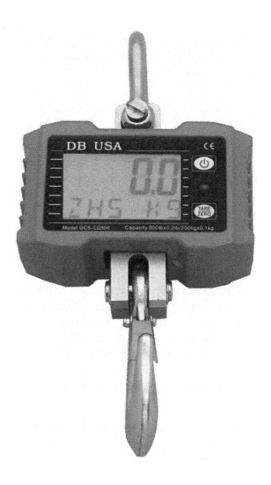

FIGURE 4.3
Chop Scales.

Industrial Scales), the lamination areas would weigh the chop strand used; and the pulling station would weigh the completed parts.

The last item to address was to determine who we would use for the time studies, as this would require the most time from regular production. We had one individual who was on light duty with a minor back issue, and was sorting screws to keep busy. This individual was also attending college at night to be a nurse, so he had the aptitude to think and work independently, so we just needed to be certain he had the desire to document accurately.

After explaining the situation to him and helping him understand how important this position was and what it entailed, we set him up with a new title as Plant Application Engineer and utilized him to provide the accurate documentation for the materials and labor needed to manufacture the hatches. He also worked with the Engineering team to correct the Bills. Consequently, his back issue was never brought up again, and he worked two years for me correcting other issues in the plant until he earned his degree in nursing.

Materials

How materials are maintained and handled is your next area of concern. Does the materials group handle all of the materials for your facility? Or will you follow a Kanban system? Will you be in charge of your own shop supplies? Will you maintain your inventories, or will the responsibility be shared with the materials group?

Having the correct materials at the correct time is essential for a smooth flow in production. Late parts and supplies create problems in flow, which in turn hurt efficiencies, quality, sales, and possibly employee retention. Get to know the purchasing department and learn the basics of their function. Generally speaking, Purchasing is concerned mostly with price, not cost as often they are not familiar with the production issues including labor to use the materials. Read an example of how a change in materials affected the cost, quality, and labor in Example 4:2.

Example 4:2   Lamination Core Material Issue.

One of the largest sport boat manufacturers in the late 90s, which was also credited with the advent of the Deck Core Kit, made a material change to go from a Deck Core Kit to a 5' square wooden box of 4" marine grade plywood blocks from a company located in Michigan. The purchasing department made the mistake of looking only at the price of kits versus the price of blocks, and failed to look at the cost of labor, quality, warrantee, and waste.

Although it was not instantly obvious, the issues from this decision included stress cracks on the sides of the cockpit floors and top radii of the sides of the decks, delaminated cockpit floors, and frustrated or disheartened lamination and gelcoat repair workers.

By the time this issue reached a head, the senior management team called all managers to a mandatory end of every day meeting to discuss lamination issues. I was managing the upholstery plant at this time, and finally had it operating efficiently, so I was not too excited about spending additional time working on issues in other plants. I was also an ISO auditor for the company and had determined that the company had neglected critical areas in the lamination department including quality assurance and chemical testing. The general moral of the employees in the lamination area was at the lowest level I had ever seen, and it was obvious by their remarks, they had lost faith in the ability of the management team to successfully run this area.

So, I'm sure I probably had a little chip on my shoulder sitting in a meeting talking about cracking issues in a plant where it appeared to me the main issue was the management team itself. And so, when prompted, I asked if anyone had performed a burn test on the laminate. A burn test measures the ratio of fiberglass to resin in a core sample usually taken from a cut-out on the part. Since there is very little strength in resin, too much of it will make the part brittle, and can crack. I already knew that a burn test had not been performed since I had previously found the lab tools and equipment covered in dust, and the lab was being used as a storage area when I performed my last ISO audit. The answer was no, in fact, the senior management team did not know what a burn test was, and I later found myself working in lamination helping to correct the cracking issues.

After observed the deck lamination process and speaking with the employees, I consulted with the Naval architect on core placement and stress points, and determined that the wood blocks did not allow the employees to correctly core the decks resulting in stress cracks near most radii.

An additional observation of the deck lamination process high-lighted the floor delamination issue, in that the box of wood blocks was delivered to the outside alley way near the deck lamination area, which was exposed to the elements including rain on occasion. Water and resin do not mix, and the wet blocks would not allow the laminates to adhere properly, resulting in delamination.

After evaluating four choices, i.e., existing wood blocks, Baltek, Diab, & SprayCore, it was determined that the company would move back to Baltek foam core kits. The cracking and delamination issues were resolved.

You may also find it beneficial to hire a material coordinator for your facility if you determine that materials are already a problem. This person can also be responsible for the allocation of shop supplies and perishable tools including drill bits and hole saws, they can track spending by employee, and help to head off issues with theft and waste. Read Example 4:3 to understand how managing shop supplies accurately can help with flow, waste, and efficiencies.

Example 4:3   Small Parts Material Coordinator

The Problem   Although it was locked when not in use, the Shop Supply storage area, or as commonly called "the Crib", was in total disarray, and there were no controls on costs. Supervision and a few crew leaders had a key to access the crib but would frequently hand the key off to the workers to get what they needed. Shop supplies at the plant included all perishable items used to produce the end product that did not actually become part of the end product. Expensive diamond drill bits and burrs would leave the shelf without notice; safety glasses and other personal protective equipment such as protective paper suits, hair socks, and gloves were always in demand and often scattered. Paper cups, tongue depressors, sandpaper rolls, and tape left the crib almost as soon as they arrived and were hoarded in employees' carts. And to make matters worse, the budgeted amount (which was a macro number) for shop supplies at this plant was $12,000, but according to accounting, we ranged over the previous six months from $9,624 to $17,402, with an average of $14,213. See Figure 4.4 for an example.

It was almost impossible to hold anyone or any department in the plant accountable to the mess or the cost since no one was responsible for the area or knew how much of which supplies were allowed to be used. We needed to get shop supply spending and accountability under control.

The Solution   You will have a hard time changing what you do not measure, so the first step in getting this area under control was to measure what we believed we needed and compare this to our budgeted amount. We accomplished this by understanding how much of which

FIGURE 4.4
Shop Supply Spending.

supplies were needed in each area by employee in one week. To do this, we had the crew leaders to determine which supplies were used by each employee working for them, and how much was needed weekly by each employee. We put this information in one spreadsheet and totaled the sums of each department by product. We were astonished to find that the expected totals were 25% less than what we were budgeted, and what we were using. This gave us a goal; we now needed to develop a plan to achieve it. See an example of how this spreadsheet was set up in Figure 4.5.

To start, a position of Material Coordinator was created for the plant to put together weekly kits by department for the crew leaders based off of what the leads documented as needing; stock the supplies received from the main stockroom; keep the "crib" in order; be the keeper of the keys in order to track exceptions when they occurred; and report to the plant management on any issues. This position was filled by an existing lead, who was also still responsible for her crew. She was allowed to spend each Friday working as the Material Coordinator, which eventually only required 4 hours. An employee

Shop Supply Weekly kit					
Crew #1	**Andrea**	**Crew #3**	**Brian**	**Crew #5**	**Bob**
2" Tape roll	5	Paper Suits each	23	Grinding Discs 24	45
Aprons each	5	Aprons each	4	Paper Suits each	25
Cleaning agent can	5	Cleaning agent can	2	Disposable Gloves pair	25
Disposable Gloves pair	25	Disposable Gloves pair	43	Spray socks each	25
Face Mask each	25	Duct tape roll each	4	Duct tape roll each	6
Masking paper roll	5	Face Mask each	19	Face Mask each	25
Paint cups roll	5	Fiber Discs 24 each	20	Rags each	2
Paper Suits each	25	Spray socks each	23	Cleaning agent can	6
Rags box	5			Drill bits, 1/16 each	45
Sandpaper Disc Roll 220	10			Saw blades	45
Spray Glue each	5				
Tongue depressors box	5	**Crew #4**	**Tom**		
		Spray Glue each	3		
Crew #2	**John**	Paper Suits each	15		
Grinding Discs 24	60	Disposable Gloves pair	15		
Cleaning agent can	6	2" Tape roll	3	**Crew #6**	**Kevin**
Disposable Gloves pair	30	Sandpaper Disc Roll 220	10	Paper Suits each	15
Drill bits, 1/16 each	60	Face Mask each	15	Disposable Gloves pair	15
Duct tape roll each	6	Rags box	3	Fiber Discs 24 each	15
Face Mask each	30	Cleaning agent can	3	Duct tape roll each	4
Paper Suits each	30	Aprons each	3	Face Mask each	15
Rags each	2	Tongue depressors box	3	Spray socks each	10
Saw blades	60	Paint cups roll	3	Cleaning agent can	2
Spray socks each	30	Masking paper roll	3	Aprons each	4

FIGURE 4.5
Shop Supply Area Breakdown.

working in her crew became a lead in training and was selected to cover her area when she was working the crib.

Next, we checked every cabinet and locker for hoarding, and brought all of the excess supplies back into the crib. This would certainly help us the first few weeks with our spending and would allow us to determine if adjustments needed to be made. It was amazing what we found. A few of the supplies did not have to be ordered for over 3 months!

Finally, we would continue to educate all employees on the costs of shop supply items and spending. We did this by putting large price tags on shop supply items, and hung them on the inside of the crib, which was nothing more than a fenced-in room on the shop floor beneath the manager's office. We also constructed a large thermometer with incremental marks of $3,000 from $0.00 to $12,000 (not $18,000) with $9,000 bolded and highlighted as the goal, and $12,000 as the cap, remembering that it was the combined effort of the leads and employees that determined what they needed.

The total spending for each week and month was also a focus point in the plant weekly meeting. And an explanation of why we needed to treat the supplies as if we were paying for them was given to help the employees want to achieve the goal. The meetings were always kept on a positive note.

The Result We easily beat our goal the first month due to the hoarding issue. The next few months were a little more of a challenge, but we did meet our goal of under $9,000 by the fourth month, and there were very few exceptions for opening the supply cabinet. We actually reduced cost in this area on average by 45%, a savings that represented about $60,000 annually. With spending under control, the plant manager had more credibility and leverage to purchase necessary replacement tools & equipment for the plant because it was believed that we were responsible with the resources we were given. The employees recognized that their efforts paid off for them in the form of getting new tools that helped make their jobs easier. We were also able to spend additional money on the break and bathrooms for the benefit of the employees.

Determine the Weak Links

You cannot fix everything all at once, so it will be important to understand which opportunities should be corrected first. To do this, you will need to also prioritize the issues you uncovered in Chapter 3 according to cost. Utilizing your Issue Matrix from Figure 3.7, include how much money is

Issue	Area Owner	5-Why Questions	Urgency (1-10)	Ease of Correction (1-10)	Cost to Change (1-10)	Cost of Opportunity Lost (1-10)	Total Score	Labor Hours spent	Annual Aproximate Cost of Labor @$35 Std Hr	Materials Used	Annual Approximate Cost of Materials Used	Total Cost to Plant
Excessive WIP - 60K	Ronnie	1. Why is there excessive WIP? To supply the main plant. 2. Why does the main plant need excessive parts? Damage or lost. 3. Why are the parts damaged or lost? No accountability. 4. Why is there no accountability? No formal scheduling system in small parts. 5 Why is there no formal scheduling system? None has been created.	10	3	1	8	22	2 hours daily spent locating lost or replacement parts. Time spent manuevering around stored parts.	$18,200	Lost Parts. Damaged Parts: 2 per day average	$208,000	$226,200

FIGURE 4.6
Issue analysis

lost and/or spent on a daily, weekly, or monthly basis as a result of each issue. Remember to include the cost of lost time and/or spent time, materials, impact to other departments or customers, costs to correct, and lost sales. See Figure 4.6 for an example.

Once you have determined in order which issues represent the most cost, you will want to understand why they exist. You will almost always find that these answers are a direct result from inconsistencies and deviations within your people, processes, or the materials you use.

Conclusion

You should now have a good idea if your employees understand their functions, and are equipped to successfully fulfill their responsibilities; if processes and procedures represent an accurate picture of what it takes to produce the product; and if the handling of materials is supportive of composite manufacturing in your plant. We can now use this information and data from Chapter 3 to help get organized for implementing our plan in Section 2.

Terms

Bottlenecks
Build Books
Burn test
Chop strand
Cockpit floors
Construction manuals
Core the decks
Crew leaders
Deck
Deck Core Kit
Deficiency
Delamination
Hatch
Hoarding
ISO auditor
Kanban
Macro number
Matrix
Organizational skills
Pulling station
Staging

Stress cracks
Time study
Variance

References

www.amazon.com/dp/B078SJ1MP7/ref=sspa_dk_detail_4?psc=1&pd_rd_i=
B078SJ1MP7

Section 2

Getting Organized

5

Know Where You Want to Be

Introduction

Most folks wouldn't plan a vacation without first knowing where they are going. The same is true if you intend to reduce operational costs in your plant. You need to understand exactly where you are as discussed in the previous chapters, and exactly where you need to be. These goals should be given to you as expectations for performing your responsibilities, or you will need to determine this if not.

Goals are not something that you want to have a vague or unrealistic idea about, and they also need to be attainable. For example, if you were taking your family on vacation, you probably wouldn't make climbing Mount Everest a goal unless everyone in your family had mountain climbing training and experience; were healthy enough to make the climb; had all the necessary tools and equipment to achieve the goal; and had the mental characteristics to see it through. You more than likely would also get help and advice from the guides and experts who climb Mount Everest on a regular basis. The same is true for setting goals in your plant. You should know your limitations with capital, employee knowledge, and experience, and have all the necessary tools, equipment, and skillsets to attain the goal, or you will need to incorporate these items into the plan to achieve your goal.

Divide your goals into two categories: Short-term goals and long-term goals.

Short-Term Goals

Short-term goals in manufacturing tend to be more organizational, are considered easier to attain, and can be worked on immediately. They typically do not require capital, and often do not interrupt production processes to achieve. A good example of short-term goals would be utilizing the 5S principles to clean and organize your facility. A well organized and clean environment will help your employees perform their function in the most organized and efficient way possible, will help to increase the quality of the product produced, and will make a statement to everyone working in or with your plant that the management team and employees care about their facility, their work, and their product. Read Example 5:1 to understand how a short-term goal helped improve efficiencies and work attitude.

Example 5:1 Upholstery Organization

I found out just how important cleanliness and organization was and the impact it had on the plant when I was managing the upholstery plant at a large manufacturer of fiberglass pleasure boats located in Sarasota, FL.

I was given an employee, whom I will refer to as Mike, who had previously been working in the Fabrication Plant, where he had cut off his second thumb. The company did not want to let Mike go, and felt he would be safer working in upholstery, or possibly they thought I would whine the least about taking him.

There were three departments in the Upholstery plant: Cutting, Sewing, and Tack. All of which required the use of thumbs and or skill. The Cutting room had several large machines and equipment used to cut and or slice rolled materials as well as smaller machines used to do the same, so this was not an option given Mike's past experiences.

The sewing room required skill as a seamstress, and the ability to maneuver small pieces of materials through a sewing machine, a skill Mike neither possessed nor would be capable of without thumbs. The tack room also required skill and the use of thumbs to pull the materials and smooth out wrinkles.

I was only a few months into managing this plant and had realized that although there were several facility items needing maintenance attention, it was difficult to get the maintenance personnel to address the concerns in a timely matter, or to have the technical skill to perform the maintenance required. Hence, the smaller maintenance items like minor repairs to the building, painting, garbage disposal, and bathroom cleaning were left to the plant to perform. I decided that my new employee without thumbs would be a possible fit here, and so I gave him some simple instructions to just fix, clean, paint, empty waste containers, maintain things, and let me know if he needed anything.

I certainly did not expect the result that this employee was going to give me. He requested paint, I gave it to him. He requested tools, I gave them to him. He requested supplies, I gave them to him. In one month, our plant started to change. The wastebaskets at each station were being emptied daily, and more often if required. The workers were utilizing this employee as an additional resource for addressing small maintenance concerns that had been previously either over-looked or neglected, and their attitudes concerning the plant they worked in started to change in a positive way.

But the big change happened when Mike decided to paint the airlines in the ceiling space above the tack employee workstations, and he decided to use multiple colors to do this. In less than one week, the first impression anyone walking into the upholstery plant received was that of awe, a positive colorful change to ordinary work, and it was instrumental in the attitudes, quality, cleanliness, and efficiencies in this area, which all started to immediately improve, and stayed that way throughout the entire time I managed this plant.

The lesson for me was to allow this to happen, let the employees own it, and let them take the credit for it. Normally, I would have preferred to have had the airlines all painted the same color, but I am sure as I reflect back that this would probably not have given us the same results.

Long-Term Goals

Long-term goals generally take at least 6 months to accomplish, and often take over a year. They can involve re-designing the workstations and work flow, and can include major capital to acquire equipment and facility changes. Regardless, they still need to be documented, and addressed with all other departments whose support is also required for their success, and put into the goals and objectives if determined that an acceptable return on investment (ROI) is achievable.

An example of a long-term goal is explained in more detail in Example 5:2.

Example 5:2 Vendor Parts

While managing the small parts plant at a large manufacture of pleasure boats and yachts, our efficiencies were negatively impacted on a continual basis by having to perform gelcoat repairs on purchased fiberglass reinforced plastic (FRP) small part product including helms, consoles, and caps. We were allowed to write off the time used to perform the work as it was charged back to the vendor, but this did nothing for the cost of opportunity lost. In short, the labor hours we were using to work on vendor product was taking us away from the work we needed to complete in the plant. This resulted in overtime, tired workers, accidents, quality issues, and had a negative impact on moral.

We had three choices: 1) hold the vendor to the stated quality standard, 2) hire more workers to perform repairs on the vendor product, 3) bring the product back in house.

The first choice would appear to make the most sense but would require inspecting every part received and rejecting failures. Because

Option 2: Hire more workers		1 extra employee @ $50 standard hr wage 50 wks Cost of materials includes Repair only		
Option 3: Bring the product back in-house		6 extra employees @ $50 standard hr wage 50 wks Cost of materials includes FRP and Repair		

		Option 2		Option 3
		1 employee		6 employees
Labor	Repair	$ 100,000.00	Tooling/Gel	$ 100,000.00
			Lamination	$ 200,000.00
			Trim/Grind	$ 100,000.00
			Repair	$ 200,000.00
Materials: 10 parts per day/250 days	Repair materials only	$ 533.00	FRP/MRO/Repair Mtls	$ 503,200.00
Cost of Vendor Product	2500 Parts	$ 1,250,000.00		
Total Option Cost		**$ 1,350,533.00**		**$1,103,200.00**
Cost Savings Positive figures represent savings				**$247,333.00**

FIGURE 5.1
Vendor Cost Analysis.

parts were ordered according to hull number and to arrive just-in-time for production in order to alleviate damage and minimize changes, a rejection would certainly slow down or stop the production line, so rejection would not work for production.

The second and third choices would require a cost analysis to be certain the correct choice is the most beneficial for the company. This was not too difficult to calculate and proved that bringing the vendor product back in-house was more beneficial than outsourcing. Obviously, the type of work was taken into consideration since this had been an operation performed previously at this plant, and there were no additional costs for equipment. See Figure 5.1 for an example of a basic total cost analysis. The Standard Hour Labor rate used includes the impact of overhead.

Conclusion

Most folks find it easy to locate and point out defects especially when it is not something they have been involved in or have been responsible for. Likewise, it is relatively easy to see when things aren't working as they should be or are as clean and as organized as possible, although this becomes harder to do when you are the person responsible for how the plant has been functioning for several years. It takes a humble manager to look at the state of affairs in their plant and see the shortcomings for what they really are: lost revenue.

Terms

5S principles
Cost analysis
Organizational
Overhead
Overtime
Seamstress
Work flow
Workstations

References

www.nytimes.com/2017/12/18/sports/climb-mount-everest.html
www.mounteverest.net/expguide/route.htm
www.5stoday.com/what-is-5s/

6

Developing the Plan to Achieve Your Goals

Introduction

At this point, we have determined what the performance standards are in each area of accountability, how we measure up to these performance standards, which individuals on our team will lead the change in each area, and what the expected timeframe is for achieving our Goals. We (as a team) now need to put our data into either a project software program or graph it on the plant conference room walls in order to develop a detailed plan with assigned dates and responsible names by subcategories of each goal in the plan in order to have the best chance of success for achieving our goals as planned.

While the management and leadership team will be assigned the goals, the employees will also need to be included in the work. It is important that everyone is working toward the same goal and understands why the change is important to their employment and the company.

Start first by listing the Goals and their subcategories.

List the Goals

Obviously, the main goals in any manufacturing environment will always be the following:

1. Zero lost time accidents
2. 0.00% Warrantee or customer complaints
3. 100% on-time deliveries
4. 0.00 Material variance
5. 100% Efficiency
6. 0% Employee turnaround

You need to always keep in mind though that most if not all owners have one goal: TO MAKE A PROFIT – MONEY … . PERIOD.

All of the issues you have uncovered at this point fall under one of the above main goal headings. Start by putting together a spreadsheet or time line which includes all of the above points and list the issues you have uncovered under one of these headings.

For example: If in your walk around you uncovered three broken ceiling tiles in your lamination area, which causes rain water to drip onto laminated parts or materials, this issue should be under the heading of either material variance, efficiencies, or warrantee depending on which area experiences the largest cost issue.

As you input the objectives to complete your goals, include start dates based off of resources. This will help to determine what your completion date for the goal will be. See an example in Figure 6.1.

The Organizational Chart

The organizational chart is an important part of your process. In Chapter 4, you should have put together a responsibility matrix for your leadership team. You will need this available to complete the organizational chart. You or someone on your team will need to understand what each employee's skillsets are in order to assign names to the goals you have listed in your plan. To develop an organizational chart, you will start with a chart devoid of names, and fill in the names according to skill sets and responsibilities. Figure 6.2 is an example of an organizational chart in title only.

You should use the following format when considering how many supervisors you will need.

Broken plant: 1 Supervisor to every 25–30 employees; 1 Lead to every 5–6 employees

Fair to good plant: 1 Supervisor to every 30–35 employees; 1 Lead to every 6–7 employees

And when you have your plant working in excellent mode: 1 Supervisor to every 40–50 employees; 1 Lead to every 10 employees.

Assign Responsibilities to the Key Positions

Start filling in the names according to job responsibilities. You will also want to define the responsibilities for each position. You can use the responsibility matrix as a guide. Make sure as you include more information in your chart, keep in mind the following:

1. Balance the workload between the team

Issue	Area Owner	Total Urgency Score	Total Cost to Plant	Goal: Zero Lost time accidents	Goal: 0.00% warrantee or customer complaints	100% on time deliveries	$0.00 Material Variance	100% Efficiency	0% Employee turnaround	START DATE
Excessive WIP - 60K	Ronnie	22	$226,200			X				
Materials are not delivered to a holding station	Sally	10	$18,200					X		
Excessive chop and fiberglass waste on the floor during lay up process	Ronnie	13	$4,560	X						
Several cutting jigs and fixtures do not exist or need replacement	Ernie	14	$10,100		X					
Ceiling lights are broken in Lay-up	Marty	16	$1,350	X						
Production scheduling system does not exist	Betsy	24	$40,140			X				
Shop Supplies are handed out when requested	Sally	29	$45,100				X			
Formal QA chemical testing does not exist	Stan	14	$22,000		X					
No calibration log books for chop guns	Marty	16	$22,000		X					
Quality defects average 10 per unit	Stan	24	$11,100			X				

FIGURE 6.1
Goal Time Line

Plant Manager

Lamination Supervisor	Maintenance & Safety Lead	Assembly Supervisor	Quality
Lead- Molds & Gelcoat Lead- Lay Up - Consoles & Hardtops Lead- Lay Up - Stringers & Buckets Lead- Lay Up - Seat Bases & Hatches Lead- Foam & Pull Lead- Trim & Grind	Electrician Waste Technician Mechanic	Supervisor - Sub Assembly & Repair Lead - Patch: All FRP Parts Lead- Console Assembly Lead- Hatch Assembly Lead- Hardtop Assembly & Shipping Lead- Seat Base Assembly Lead - Shipping	Inspector Driver

FIGURE 6.2
Organizational Chart Titles Only

2. Have the appropriate number of leaders per employees
3. Have specific responsibilities for each job function with goals assigned as you outlined at the beginning of this chapter.

Utilizing software can be very beneficial for understanding resources, tracking progress, and keeping tasks on track with completing the goal. If you are familiar with a project software such as Microsoft Project, you can make a Gantt chart to help determine when projects will complete based off of start time and resources. You can adjust completions by allocating additional resources to the project. For more information and help, search online for "free project Gantt charts".

Conclusion

We now have a detailed plan to achieve our goals and have organized this data either in a project software program or graphed on the plant conference room walls. We have gained the agreement from all of the leadership team in place to achieve the goals we have set and are almost ready to start implementing our plan. All that we need to do is get our plant in a clean condition to make a statement to our employees, which we will discuss in Chapter 7, and to have the greatest chance of success with our plan.

Terms

Employee turnaround
Lost time accidents

Material variance
Performance standards
Project Software
Resources
Responsibility matrix
Skillsets

7

Clean It Up

Introduction

Before we start working on our plan to achieve our goals, we need to get the cleanliness and organizational aspect of our plant in order, and set some priorities for keeping it that way.

There are several beneficial reasons for this.

1. It will make a statement to our employees that we actually care about the facility and the work that we perform, which will in turn help to instill in them a sense of pride in the company they work for, and the work they do.
2. It will make a statement to our boss/owner that we are responsible and actually care about the facility, the people working in it, and the quality of our product. This in turn will help us get additional funds as needed to complete some of the issues that we uncovered in Chapter 2.
3. Employees will spend less time looking for tools and supplies to get their function completed, which in turn will keep them at their work center longer and allow them to complete more work.
4. When tools and supplies are located in the appropriate place that makes it the most beneficial for completing a task, there are less lost time accidents.

We do not need a Gantt chart to understand how to clean and organize our work centers, how to dispose of waste and empty waste containers, how to keep safety areas clean and contained, and how to ensure that our employees adhere to best practices concerning keeping their work space clean.

Recognizing the Difference

Some folks are organized by nature, it's just how they're made, and how they live their life: everything has a space and needs to be in it or be in use.

And, I have met others that if you looked at their desk you might wonder how on earth are they able to get anything done with all that mess? However, many of them know exactly where each piece of paper is and what is on it. If you tried to organize their desk, they would probably have a meltdown. I have always been strong on the side of the organizational characteristics, as I have found this to help produce the best results in the least amount of time.

If you were the only person working in your facility, having a disorganized work atmosphere might not be such a detriment to production, but when the plant has multiple people working at the same time (which is usually the case), having a disorganized or unclean working atmosphere is at best a hindrance to efficiencies, and almost always one of the root causes of quality, efficiency, and safety issues.

There are many different types of personalities, but you can and should expect that your employees adhere to your policy on plant cleanliness and organization.

The Blame Game

Too often in my travels, I've heard employees and management blame the cleanliness of the plant on one or two reasons:

1. We have tried to get our employees to adhere to clean practices, but they either can't get it or don't want to.
2. The type of work we do is not conducive to a clean facility.

Of course, these are both excuses, and their root cause is inexperienced or inadequate management.

Concerning Point 1 above: As an example, if you had three children, and two were clean and organized, would you accept that one should not have to clean up behind him/herself? Would you do the work for them? How long would this last before the other two children cried mutiny? Humans are always concerned with fairness especially if they feel the scale is tipped against them. If you accept less than adequate work from your employees, that is probably what you will always get. A great example can be found in Example 7:1.

Example 7:1 Upholstery Bathroom Detail
The Problem: As a new manager of 90 employees in the Upholstery plant of a large boat manufacturer, I was horrified at the cleanliness of the Ladies bathroom. There were cardboard boxes located behind each toilet, which looked to contain feces, and the stall doors had graffiti with disrespectful verbiage all over them. It was disgusting, at least from my perspective. After inquiring from our plant material coordinator, why and

who had placed the boxes in the stalls, I was told that if we did not do this, the ladies would throw their wrapped feces on the floor.

It would have been easy to just make a statement and post it to the doors, but I decided that we had several issues at play here. 1. We had some cultural issues, which we needed to delicately handle: several of our ladies were from a 3rd world country that did not have a sophisticated pluming infrastructure like we had in the USA. These ladies considered it a gross offense to dispose of feces in someone else's toilet. 2. We had some non-cultural cleanliness issues that we also needed to handle. 3. We had a respect issue that we needed to handle: one of our ladies had an issue with the former management and had brought up an Equal Employment Opportunity Commission (EEOC) charge against the company. To say this individual had a chip on her shoulder would be an understatement. Much of the verbiage was more than likely a result of not handling this issue correctly.

The Solution: I called a plant meeting in the air-conditioned cutting room. All 90 employees attended, which included 15 men. I was careful to not specify which bathroom but made it a joint issue. I let the employees know that I understood the cultural differences and helped them understand that in the United States we consider it a gross offense to not dispose of feces in the toilet. I explained that we would be painting the bathrooms, and would expect that all employees would be responsible for cleaning up after themselves, and if they used the restroom after another employee, and found the stall unacceptable, to let management know. I also instructed them that graffiti would be considered destruction of company property and would not be tolerated. I heard a loud comment from the woman who had the company on an EEOC charge state that she wasn't going to sit on the toilet, she would step on the seat and expel. To which I replied that I really didn't care how she did her business, and I certainly would not be following in after her to check. She was welcome to do cartwheels during the process if she liked, but when she was finished, she had better clean it up.

The Result I: instructed our material coordinator to remove the card-board boxes from the bathrooms. We never had another issue with our bathrooms not being clean.

Concerning Point 2 above: I don't know of a composite manufacturer that does not produce some dust. Even with vacuum-assisted devices, there is still the opportunity for dust when the tool breaks the suction on curves and radii. And yet many composite manufacturers still practice clean work habits and are successful at keeping their facility clean and organized. It is a matter of policy, and when the expectation is to clean as you work, the end result is higher quality and higher efficiencies. Another great example of how working clean in a dusty environment will increase efficiencies and quality and will reduce labor can be found in Example 7:2.

Example 7:2 Repair Best Practices
The Problem: The gelcoat repair depart of any composite account is generally a dusty environment to work in. In a former position as

a powerboat plant manager, I found myself generally frustrated with this department due to the inability to keep them on a consistent flow. Gelcoat defects and time to repair fluctuated with the level of quality coming from the lamination department, and with repair technician expertise.

The Solution: It wasn't until I took a position with Norton, a brand of Saint-Gobain, and an abrasive manufacturer selling abrasives including sandpaper through distribution to the marine manufacturing industry that I realized the impact of keeping the abrasive clean while sanding. The grains that are hardly felt on the sandpaper are like rocks that fracture when pressure is applied. These fractured pieces combine with the dust to create contamination on the disc, called swarf. The swarf takes the shape of small dots, and act like a much larger grain size. If the employee does not keep the sandpaper clean, they will produce deeper scratches which may be impossible to remove. Consequently, they spend on average between 15% and 25% more labor and buffing compound to take out the scratches that should never have been put in to begin with.

The Result: After educating my customers on how to sand properly, and how to keep the sandpaper clean while sanding, my customers have saved more on labor than they actually spend on the abrasives they use (generally about 5k per employee using abrasives), and this does not include the increases in throughput and quality. Just because an environment is dusty, doesn't mean you shouldn't practice clean workmanship.

Conclusion

When it comes to cleanliness and organization, accept no excuses. Make sure you get your facility in order concerning cleanliness and organization before you start working on your plan. This should not take longer than a few weeks, and all employees should help.

Terms

Buffing compound
Cutting room
EEOC charge
Gantt chart
Swarf
Throughput
Vacuum-assisted device

8

Work the Plan

Introduction

Working the plan is often the most challenging part of realizing your goal for reducing costs. As mentioned previously, it is not too difficult to find the things that need improvement. The real test of patience and persistence to realize the goal is in working the plan. Begin by teaching the basics of how to report progress and what to do when things fail. Bring the entire facility in on the changes and help them understand what the challenges are. The best results will happen when everyone stays informed on the progress. Be open to adjusting the plan if absolutely necessary or if adjustments will produce more favorable results. Finally, celebrate the desired results at every opportunity.

Teach the Basics to the Team

Ok, so you have researched, set goals, outlined a plan, and picked the team. Now you need to equip the team with the tools needed to win.

Start by having individual meetings with each team player daily for a period of at least one week to go over what the daily goal is, how it is going to get accomplished, what challenges have surfaced, what solutions for the problems, and what progress did you experience on the previous days goals.

This will set a pattern and will help to avoid embarrassment once the weekly meetings start with the group. Meetings are expensive and lock valuable management time, so make sure the leaders understand in advance what questions will be asked and how to answer them. An example of an excerpt from a good meeting is as follows:

MANAGER: "Ok Bob, did you accomplish the goals for yesterday?"

BOB: "All but point number 3, boss. We damaged a piece of upholstery and were unable to get it repaired in time for shipment. We are down 1 order. We expect to make it up early this am and will be on track by 10:00am."

MANAGER: "Do you foresee any problems with today's goal?"

BOB: "None, we will be on track at the end of the shift."
MANAGER: "Great job Bob. Thank you."

You should always target 15 minutes for a meeting and no more. To accomplish this, everyone needs to understand three basic points and be prepared: 1. What the meeting is for; 2. What will be discussed; and 3. What answers are expected of him or her. It is your responsibility to keep the focus on the agenda, and not let the conversations divert from the agenda.

Also, if you notice one of your direct reports floundering, help them learn how to get the job completed on their own. Show them what to do. Educate them. If done correctly, it is much more beneficial to train an existing employee than to look for, hire, and train a new one.

Keeping Everyone Informed on the Progress

If possible, utilize the break or common areas to help the employees stay informed on the progress of the plan. Plant meetings work well for this and can be done by area if work schedules prevent the entire plant from meeting at the same time. Be sure to be present at as many of the meetings as possible.

Some employees are reluctant to change, and I have found that this is generally due to a lack of understanding for the change. Help your employees feel secure that the changes are in their best interest as well as the companies. A financially stronger company can invest in better tools and equipment for its employees. An efficient plant is more likely to get faster approvals on Capex.

When I managed the small part plant for a large marine manufacturer, we had an issue with the cost of maintenance of our chopper guns. Read Example: 8:1 to understand how information can be beneficial when presented properly.

Example 8:1 Chopper Gun Maintenance Issue

The Problem: A small part plant for a large boat manufacturer in Sarasota, FL had 11 total guns: 7 were chop, 3 were gel, and 1 was for putty. The average cost to maintain these guns at this time was $750 per month. Gun maintenance was new to this facility, which had recently hired a new maintenance lead. Up to this point, the main plant, which was located a few blocks away, was responsible for attending to all of the maintenance issues at the offsite small part plant as well. This did not always happen as needed, and so we hired a maintenance lead disguised as a hardtop assembly lead. I will refer to this gentleman as Bill.

The Solution: I instructed Bill to track spending on our guns and equipment in an effort to get a baseline for each gun. We then assigned an employee name to each gun on the spreadsheet, and Bill tracked all charges spent on the guns by employee.

As previously mentioned, on average, we were spending around $750 per gun monthly and were ready to post the results on a big poster board in the plant break room for all to see. However, I needed to take the opportunity to congratulate the employee with the least amount of cost per gun in front of the entire group first. It wasn't as important that all of the guns at this point were costing too much to maintain, I was only trying to motivate the gun operators to reduce their individual spend. I made it a point to thank the gun operator with the least cost of maintenance in front of the entire plant and let him know that I valued his ability to care for his gun.

The Result: Maintenance on all of the guns was reduced on average to around $30 per month. Bill helped with these results by working with the employees on proper care and calibration of the guns.

Adjust Only as Necessary

Adjustments to the plan should only be made when resources dictate that the plan will not be achievable. For example, fluctuation in the production schedule may require additional resources that might need to be taken from the plan. One way to combat normal issues with production while working your plan is to require the supervision to bring solutions when they discover issues that present themselves as stumbling blocks to achieving the goals. It never ceases to amaze me of what the human capacity is regarding work-load. Don't overwork your employees, but don't hold them back either.

Celebrate the Desired Results

Make sure you are tracking the progress of each of the direct reports on a chart, which can be posted for all to see. Always publicly praise the direct reports whose results are desired. This will prompt for future behavior of the same. It is ok to discuss inefficiencies as well, as long as it is not done in a disrespectful or negative nature. The team may have comments to share on how to improve the progress of a flailing area. Just be careful to keep the tone positive. Negativity breeds negativity; positive attitudes produce positive results.

Conclusion

Accurate and current information are necessary when implementing your plan to reduce costs in the plant. It is important to keep everyone informed of the progress and the issues. Teaching the leaders to do this with their teams is important to help them stay focused on the goals. Adjustments may need to be made to the plan, but only as required, and don't forget to celebrate every win, even if it is just by sharing the good news. This will keep the team motivated to continue when the long-term goals wear them down.

Terms

Calibrate
CAPEX
Chopper gun
Hardtop
Human capacity
Inefficiencies
Production schedule
Workload

References

www.investopedia.com/terms/c/capitalexpenditure.asp

Section 3

Hiring and Keeping the Best Employees

9

Character Skills vs. Experience

Introduction

Choosing the best person for the position is often left up to the joint decision between Human Resources and the Plant management team in the hope that the applicant will be vetted and found to be free from illegal drugs, crimes, and other unethical, illegal, or immoral behaviors; and will possess the necessary qualifications to perform the position appropriately. But I have found that too often management settles for the experience without paying attention to the personality characteristics of the individual, or even worse they settle for no experience and no consideration for the character of an individual. It often becomes a matter of needing a body, and the thought that prevails is "*we have to get someone in here now*".

The composite industry is still predominantly a labor-intensive industry, and so the selection of individuals who are willing to perform this type of work are often few and far between. In many companies, the employee base for this industry is mainly immigrants who often do not speak English. This presents an additional challenge when trying to determine if an individual possesses the desired character skills to be a productive long-term employee.

To sum it up, the questions that need answering are: Why are character skills more important than experience? What are the costs for manufacturing from hiring someone with poor character skills? And, how does a manager hire the best folks when communication is a barrier?

Character Skills vs Experience

I've often heard the statement "You can't teach an old dog new tricks", or to put it in perspective: veteran employees don't change easily. But I have found there are exceptions to this rule, and it has to do with the character of the individual. To fully understand, we have to consider what types of characteristics are desirable for manufacturing, and which are not. I have categorized some of them as either Negative Characteristics or Positive Characteristics. See Figure 9.1.

Personality Characteristics	
Negative	**Positive**
Rude	Kind
Impatient	Patient
Arrogant	Humble
Stubborn	Flexible
Dishonest	Trustworthy
Disloyal	Loyal
Lazy	Diligent

FIGURE 9.1
Personality Characteristics

A person with positive characteristics will be able to look objectively at change and consider how they can work with it to better their job, their co-workers, or their company. A person with negative characteristics will automatically be looking for reasons why change will affect them negatively on a personal level, and they will generally promote this belief to their co-workers. Reference the example given in Chapter 7, Upholstery Bathroom Detail. I not only had to deal with the issue, but I had to also deal with the negative comment or allow it to permeate with the other employees. On the flip side, employees with positive characteristics take less time to manage, can be counted on to work independently, are respectful to their co-workers and management, and generally produce at a higher quality level. Read Example 9:1 to understand how great character skills surpassed experience.

Example 9.1   Maintenance Lead

The Problem: While managing the small part plant for a large marine manufacturer in Sarasota, FL, I was faced with several facility maintenance issues. The plant was offsite from the main manufacturing location but was still being maintained by the main plant. When equipment failed, I or one of my supervisors would call the maintenance office and request help. Often it was a crisis situation where entire crews were held up waiting on the maintenance tech to arrive to repair a gun, pump, heater, or agitation system. On occasion, we lost a part due to the lack of preventative maintenance and the time to repair. We were not afforded our own maintenance person, although this is what was needed.

The Solution: The plant was responsible for manufacturing and producing all small part assemblies to the main plant, and included 90 employees, 20 of whom worked in the assembly functions installing hardware and electronics onto the FRP product. After losing an employee in the hardtop assembly area, I decided that instead of hiring a replacement, I would hire a maintenance person, and title them hardtop assembly. And, I would let my boss know my intentions. I instructed my assembly supervisor to start looking for and interviewing a person with the capabilities and characteristics we needed for this position.

Applicants were selected and interviewed, but after a month we still did not have a maintenance person. After questioning my supervisor regarding one of the candidates, I was told "he did not have the qualifications to do the job". This gentleman was in his 50s, worked in maintenance at a juice processing company nearby, and had never worked in the composite industry, which was why he was not chosen. I requested he be brought back in for an interview that I would give myself.

After interviewing this gentleman, whom we will call Bill, I realized that he had all of the character skills I desired and felt certain he would pick up on the equipment and tools relative to the composite industry, and I was correct. I let Bill know that I knew very little about maintenance but believed in preventative maintenance not reactionary maintenance. I took Bill on a tour of our facility, which included broken ceiling tiles, damaged roller doors, damaged metal walls with holes punched through, lack of maintenance on all equipment, and an almost completely disorganized maintenance shop. I let Bill know that the changes that needed to be made would be a credit to him, and all he needed to do was to let me know what obstacles were in his way, and how I could help him succeed.

The Result: Bill not only got our facility on preventative maintenance, but he also became the go-to person for the main plant. In fact, they tried to take him from our facility and promote him to Supervisor over all of maintenance. He let them know he was not interested leaving his position in our plant. Bill rebuilt several (at least 7) resin pumps and had them on standby, which the main plant also used. These pumps, which were considered scrap, cost over $4,000 each when new, and only cost a few hundreds to fix.

The maintenance department grew from no employees in our plant to three, as we hired an electrician, and were given a gentleman who was considered non-functional in the main lamination facility. I added him to the maintenance team, and he became our janitor. Consequently, our plant did not have issues with maintenance that affected the normal production processes. We would have missed out on Bill if we had only looked for experienced help.

The Cost of Poor Character

There are many quotes and sayings about character, but one of my favorites that warrant mentioning is:

He that lieth down with dogs shall rise up with fleas.

Benjamin Franklin

Point number 1 gets right into the first issue of hiring an employee with poor character. When you consider that we spend at least one-third of our time with our co-workers, it makes sense that we will pick up some of each other's traits and characteristics. People learn from each other, and a person with negative characteristics will infect the good behaviors of other employees in the plant.

Point number 2. You will frustrate the employees who are performing as they should, when the work they do is dependent on the person with poor characteristics.

Point number 3. You will find yourself spending additional time managing the employee with poor character skills.

Point number 4. This employee generally does not last, so training and hiring time will be wasted.

Point number 5. If their character skills are unethical, you could put the company in a position dealing with legal suits, work compensation issues, or other. See Example 9:2 for an example.

Example 9:2Lost Safety IncentiveThe grinding booth department at a small part facility needed an additional employee to cut and grind the fiberglass consoles and helms. This plant was considered safe and had experience over 465 days with no lost time accident, which allowed them to have a catered dinner every Wednesday, and a quarterly steak dinner. The entire plant of 90 employees were very careful to work safe and not lose their free catered weekly lunch. It was a benefit that all enjoyed.When new hires were selected, the employees also helped to instruct them on safe practices. And as recommended by one of the lead workers in Lamination, the nephew of a worker was selected for the position. He was hired due to his aunt's reputation and character.After he passed his 30-day mark, he filed a work compensation complaint for a back issue, and the safety incentive we enjoyed for over a year was gone. We later heard from one of the other workers that he did this to collect pay from the company, while working under the table, but was never able to prove this.

The Communication Barrier

When communication is an issue, and it is all you have coming in the door, your best hope is to educate yourself on the cultural aspects of people you are looking to hire and take extra time interviewing for each position. You might be surprised to learn some of the cultural differences that can cause issues in manufacturing.

For example: In the plants that I have managed, I have had people from Russia, Mexico, Haiti, Dominican Republic, Barbados, Taiwan, El Salvador, Guatemala, Spain, Ukraine, Cuba, Puerto Rico, Nicaragua, Honduras, and several others, who did not speak English. While most of them are Spanish, their dialect and translations can be quite different. It is also important to know that they don't all like each other either. But the biggest lesson I learned from managing different cultures of people was an occasion in Lamination with a Haitian employee. I've always believed to reprimand in private, praise in public. Haitians (I am told) feel insulted if taken to the manager's office even for a positive review. They prefer for you to tell them right out on the plant floor.

Conclusion

Of course, the ideal employee is the one with experience and positive character skills, but if you have to make a choice, always vie on the side of character skills. Take time to learn about the employees you plan to hire as well as their culture if different from your own. It will be time well spent.

Terms

Agitation system
Character skills
Grinding booth
Human Resources
Immigrants
Labor-intensive industry
Preventative maintenance
Reactionary maintenance
Resin pumps
Roller doors

10

Motivation

Introduction

Motivating employees to perform at optimum levels is a topic that arises on a continual basis at most manufacturing facilities I visit. Since labor accounts for the majority of cost in most open-molded composite plants, it is important to understand how to motivate employees to produce a quality product safely, efficiently, and consistently in order to control the cost of labor.

How do we motivate our employees, and how do we keep our employees motivated in tough times are the standard questions most managers ask when speaking about labor?

There are three types of motivation: Positive, Negative, and Discipline, but understanding the types of motivation is putting the cart before the horse. If you want to motivate your employees and keep them motivated, you need to first know what motivates them, that is, what is most important to them? The answers may surprise you.

A Lesson from Maslow

Before we can determine what motivates our employees, we need to understand a few things about needs. An American psychologist in the 20th century, Abraham Maslow, created a famous pyramid to depict how important needs are to humans, and that they have a hierarchy or importance when dealing with people. See Figure 10.1 to gain a better understanding. Basically, what he is telling us (and this may sound like common sense) is that certain needs must be fulfilled prior to a person having the ability to be motivated. For instance, it might be hard to motivate an employee to produce more if they are dealing with a terminally sick child or spouse. The hierarchy shows that basic needs must be met before a person is in a position to be able to motivate. There are three stages: Basic, Psychological, and Self-Fulfillment. Some employees are at the Self-Fulfillment stage and can motivate themselves to

FIGURE 10.1
Maslow's Hierarchy of Needs

produce more with very little encouragement. Others in the Psychological needs category are experiencing no issues with Basic Needs and are prime candidates for motivation. But, you will find it a challenge to motivate the ones in the Basic needs category unless you can solve the basic needs issue. At the very least, you should understand your employees, and learn what category they are in and possibly what motivates them.

Positive Motivation

Of the three types of motivation, Positive Motivation will produce the best results. Positive Motivation is rewarding someone for doing something good. This is not necessarily a tangible item. I have found it more beneficial to praise employees for doing something well, than focusing on the ones who didn't do it correctly. And if you understand your employees, what character skills they have, and what motivates them, you can use this to benefit them as well as the plant. Read Example 10:1 to understand how of forward-thinking positive motivation helped to reduce cost.

Grinding Booth Cost Reduction	
Existing Situation	Hourly Rate of the Grinding Booth
7 Employees x $50 Std hr wage with overhead	$350
Total for Existing Situation if we hire a replacement	**$350**
Projected Situation	
6 employees x $50 Std hr wage with overhead	$300
6 employees x $1.00 Premium for additional work	$6
Total for Projected Situation if we do not hire a replacement	**$306**
Hourly Rate Cost Savings	$44
Projected Annual Savings for this department @ 40x52x$44	**$91,520**

FIGURE 10.2
Grinding Booth Cost Reduction

Example 10:1   Grinding Booth Reduction

The Problem: The grinding booth for a composite small part manufacturer had seven employees, called grinders. One of the employees had been with the company for 25 years and had decided to retire. Hiring and training a grinder to replace the retired employee meant that even an experienced one would go through a six-month training curve to learn how to cut the complicated parts that were manufactured at the plant.

The Solution: After getting the proper approvals, I called the entire grinding department into my office and presented the following idea to them.

> Instead of hiring another grinder and working through a six-month training curve, I will pay the six of you an additional $1.00 per hour to absorb the work of the seventh grinder. If you agree with this, you will be expected to keep up with production with no overtime. This increase in pay will not interfere with your normal review.

The Result: All six of the employees agreed wholeheartedly with the change and absorbed the work without a problem. This represented a $91,520 annual cost reduction for the department (see Figure 10.2).

Negative Motivation

Negative motivation is still motivation, but it happens by using the opposite technique from positive motivation. In positive motivation, the employee is given something to motivate them: praise, money, a promotion, etc. In negative motivation, we take something away from the employee to get the desired result. This can be best understood by reading Example 10:2.

Example 10:2   Upholstery Musical Chairs

The Problem: As a new manager to an upholstery plant that had 90 employees, I was faced with the challenge to motivate the employees to follow my lead. There were three departments that produced the upholstery used in the assembly plant for producing boats: cutting, sewing, and tack. The tack department had 36 employees all of whom were not keen on gaining a new manager that they felt did not have upholstery experience and challenged every decision I made. I put up with this behavior for a month as long as they kept up with production but grew tired of the insubordination as it was becoming a deterrent to the corrective work we needed to achieve to reduce cost, and to make the plant efficient. I knew that a silent mutiny would cripple the assembly plant, and many folks depended on the income they received from working at the boat plant. Being sent home due to lack of upholstery might eventually end my career but would certainly end the careers of many of the employees in my plant first as well as hurt the assembly plant.

The Solution: I remembered speaking with an elementary school teacher who explained how she moved desks around the room when the children get too comfortable with their peers and don't listen to her as well as they should. Why not? I thought. I will take away their comfort zone with each other, and move them to opposite locations, taking into consideration to separate the ones who feed negativity off each other. The plan was to come in on a Saturday with my maintenance employee, and basically play musical chairs with their tack tables. I was counting on the probability that they would come in on Monday morning, get nothing done for at least an hour, and be shook up enough to follow my lead.

The Result: It worked, although I'm not really sure if it worked because they thought I was crazy, might possibly fire them, or because they realized that their positions can easily be changed. By taking away their comfort zones, they had no choice but to take direction from me, and the plant thrived. Efficiencies rose from the high 40s to the high 90s, waste declined about the same as efficiencies, and a sense of pride was realized in the plant. Not all of this was due to negative motivation, but this was the turning point for this department. A perfect example of how negative motivation can help give you a positive result.

Discipline

The last resort to motivation is discipline. It is not as effective as the other two and should only be reserved for extreme cases after everything else has failed. This is sometimes confused with negative motivation, but the difference is punishment. My parents always told me to never spank my children unless they did something that either caused injury to another person or had the potential to cause injury to themselves or others, and you needed them to remember the discipline. His example was if your child crosses the street without looking both ways, you want them to remember the spanking, so they never do this again. It made sense to me, and I can tell you that I can remember all three spankings I received from my mom.

Discipline in a plant generally takes the form of dismissal. Unfortunately, it is the last resort, and, if done correctly, does motivate the other employees left behind to fall in line. Read an example for how this can be beneficial in Example 10:3.Example 10:3Steam Roller

> **The Problem**: I was helping to organize the Lamination plant at a large boat manufacturer in Sarasota, FL. The plant was experiencing low efficiencies, poor product quality, and management issues. Although most of the employees had over 20 years of experience working in this plant, their attitudes toward the company and the management team was in the toilet to say the least. I realized that the main issue with the attitudes resulted from several changes to the product, plant, and work experience of the employees. For instance:
>
> - The stockroom would consistently drop off product in locations where it was difficult to access the materials needed to manufacture the parts. On occasion, the materials were contaminated from being exposed to the elements, but there was no concession for getting replacement materials.
>
> - Production control would make no allowances for molds that needed to be removed from production and reworked or replaced. This caused several issues with gelcoat repair. Work order changes would happen after manufacturing had already completed some of the lamination process, and laminators were required to rebuild areas of the hulls in assembly.
>
> - Engineering failed to follow the product through manufacturing to be certain jigs and fixtures were accurate, and lamination was expected to rework areas in the assembly plant when components did not fit.
>
> - Quality assurance did not exist for Lamination, so all rework was completed by the laminators in the assembly plant. Raw materials were not checked during the process, so issues were dealt with as they happened.
>
> - The employees had basically lost all faith in the company to manage the lamination area, and consequently felt that if the management didn't care enough to understand and accurately manage basic lamination processes, why should they.

The Solution: After dealing with all of the above issues that were outside of the control of the employees, the only thing left to correct was the attitudes of the employees. It would be a challenge to gain in efficiencies and quality as long as the employees didn't care. Unfortunately, we did not have the time to wait on them to rise to the occasion, and decided to motivate them through discipline. We did this by hiring a personality type A, a steam roller, for lack of a better word, to manage this area. This gentleman had a history with mechanics as well as a strong background in boat manufacturing. The one thing he did not have, was experience managing lamination. But his personality was needed to quickly fire the employees up, or fire a few of them as examples. Not an ideal solution, but it worked well, and the situation was desperate.

The Result: The lamination plant thrived and functioned as a well-organized and maintained manufacturing facility. Quality and efficiencies rose to the low 90s. A few folks had to be fired to shake up the remaining employees, get them to accept the changes, and work with the new manager.

Sometimes discipline works well, but almost always at a cost to human capital. If you are faced with a similar situation, be sure to change this management style after no more than six months to prevent a loss in morale.

Conclusion

In Chapter 10, we learned that motivated employees produce higher quality products more efficiently than non-motivated employees and require less management time. Motivating employees also requires that we understand what it is that's most important to them, it also might require that we understand a little about their culture, if different from our own. Of the three types of motivation, we learned that positive motivation always works best, and should be the first tool we use when possible.

Terms

Components
Contaminated
Discipline motivation
Forward-thinking
Gelcoat repair
Hulls

Maslow's Hierarchy of Needs
Molds
Negative motivation
Open mold
Positive motivation
Stockroom

References

www.simplypsychology.org/maslow.html

11

Choosing and Empowering Leaders

Introduction

In this chapter, we will grade the direct reports in a matrix in order to be sure we have selected the correct leaders for the tasks, and according to the following: works well with others – no negative personality issues, work ethic – diligent, motivated, dependable, loyal, organized in thought and actions, knowledge and/or experience in responsibilities, and open to change. This is followed by selecting the leaders who possess the desired qualities to get the specific job completed.

Next it is important that your leaders feel supported and equipped to lead their teams, and you will want to be certain your team stays motivated to succeed. Breaking the chain of command can cause confusion at the very least, but often results in the leader having a more challenging time getting the employees to follow them. It will also consume management time needed for managing costs in the plant, and does not empower the leader to excel in their position. There is no point in having a leadership team if you don't follow the chain of command it represents.

Categorize the Existing Leaders

Take the organizational chart you developed, and grade your direct reports according to the following:

- Works well with others – no negative personality issues
- Work ethic – diligent, motivated, dependable, loyal
- Organized in thought and actions
- Knowledge and/or experience in responsibilities
- Open to change

Rate each point above for each direct report on a 1–10 scale with 10 as excellent and 1 as poor. You should have each direct report grade themselves and each other on the same scale.

Keep all information confidential. Absolutely do not share negative comments between the direct reports. This will only breed hostility. Contrary to popular belief, criticism is never constructive. It is usually only taken as criticism. You are gaining this information as a basis only. When looking at the remarks from other direct reports, you need to keep in mind that it is their perspective. While perspective can be useful, they are always based off of one's perception, which is influenced by one's beliefs, and may not necessarily be 100% accurate.

Take all of the grades and combine them to get an overall grade for each point based on your, their, and their co-worker's assessment. This information will allow you to target the weaknesses in each of your direct reports in an effort to develop them for the positions they hold. See Figure 11.1 for an example of this chart.

Assign the Leaders to Their Key Positions

Review the notes and charts on the direct reports. For the positions, which already exist, leave the leaders in place for the first month. On new positions, and based off of the charts, select the leaders who possess the desired qualities to get the specific job completed. You might also find leaders in your organization that would not normally seem to be a wise choice. You will have a decision to make: replace them or find a way to get them to want to change. Make sure you give yourself time to see the entire picture before you decide to replace an employee, especially one in a leadership position. There are a few good reasons for immediate dismissal such as unethical behavior, theft, lying, and safety issues, and you will want Human Resources involved when this becomes necessary. You should never dismiss an employee based off of personality conflicts until you have given them the opportunity to change. You might just find the ones who challenged you the most turned out to be the best leaders you've worked with. Read Example 11:1 to understand from an otherwise destructive personality situation.

Example 11:1   Follow Me

The Problem: Alice held a position of group lead in the plant, a position with five direct reports, whose responsibility was to make parts for customer service and warrantee. She had been at the company for over 20 years, had a wealth of knowledge about the different types of product we manufactured, and was considered a strong resource for warrantee

Leadership Rating Matrix

Rating	Works well with others				Work ethic				Organized				Knowledge/experience				Open to change				Composite Score	Rating
	Self	Peer	MGR		Self	Peer	MGR		Self	Peer	MGR		Self	Peer	MGR		Self	Peer	MGR			
Jack Jones	8	7	9	8.00	9	9	9	9.00	6	6	7	6.33	8	8	9	8.33	9	8	9	8.67	40.33	8.07
Jim Hat	7	8	7	7.33	9	10	10	9.67	8	7	8	7.67	8	9	9	8.67	6	7	6	6.33	39.67	7.93
Larry Peters	8	7	8	7.67	7	8	7	7.33	8	7	8	7.67	9	9	9	9.00	8	8	8	8.00	39.67	7.93
Rose Rogers	8	7	7	7.33	8	9	8	8.33	8	8	9	8.33	9	9	9	9.00	8	8	8	8.00	41.00	8.20
Don Diminic	7	8	7	7.33	9	8	9	8.67	7	7	7	7.00	9	8	9	8.67	7	7	6	6.67	38.33	7.67
Beth Barnes	9	9	8	8.67	7	7	7	7.00	9	8	7	8.00	7	7	7	7.00	8	8	8	8.00	38.67	7.73
Chris Smith	9	8	8	8.33	9	10	10	9.67	8	8	7	7.67	7	8	7	7.33	7	8	7	7.33	40.33	8.07

FIGURE 11.1
Leadership Rating Matrix

and customer service in reference to how older product was manufactured in the plant. She had one problem: she challenged every decision management made and had a negative attitude toward the company in general.

I was the new manager for this plant and had already evaluated all of the management team in place. I knew that Alice would either have to change her attitude, or she would have to be replaced. This was not a decision I wanted to rush into due to her product knowledge and experience, but how do I get her to want to change? I had already taken her into the office and had a private conversation with her on her work at the company, and how she felt about it. Her tone and body language proved she did not trust management and had little faith in the direction of the company. She clearly did not want to take direction from me. To make matters worse, she had filed a complaint with the Equal Employment Opportunity Commission (EEOC) against the previous management team, one of whom was one of my two supervisors in the plant.

The Solution: Reflecting on motivation techniques, discussed in Chapter 10, proves of the three types, positive motivation works best. I had already thanked her on occasion for the great work she was doing for customer service and received a cold thanks as if she placed no value in anything that I said. I would continue along this path for a few weeks longer, give no attention to the negativity as long as it didn't get worse, and wait for an opportunity to call her out on her attitude.

It didn't take long. She wanted attention, and control, and when she changed a process I put in place regarding an order for customer service, I let her complete the order, and pulled her into the office.

I started the conversation on a friendly note and told her that "I valued her experience and knowledge regarding our product, and viewed her as a natural leader". She did indeed possess the qualities to lead and direct others well, and they followed her. To this, she relaxed and sat back smiling in an "*I know*" way. I followed this compliment with another and told her that she had the ability to learn new challenges in our company, and I needed to have all of the warrantee product digitized for our Gerber Computer Numerical Control (CNC) cutter (a change that would greatly reduce our dependency on her). I let her know I was considering a few people in the plant for this new responsibility, and then I hit her hard (verbally). I told her again that she was definitely a leader, and followed this comment with another, in a direct and definite way, that there was only one problem with her leadership skills, and I waited to be certain she was listening.

> You are leading the team in the wrong direction, you are leading them against my decisions, and I won't tolerate it. You

need to change this immediately, before you will ever be considered for higher positions in this plant, and before you lose your position as lead.

After she left my office, I treated her the same as I always had, the same way I treated everyone else in the plant, with respect and dignity.

The Result: Alice was the best choice for the digitizer position, and she changed her attitude, was promoted, and became one of the best leads in the plant, and an even more valuable asset to the company. She dropped all EEOC charges and became good friends with the Supervisor charged. The company saved the time to look for and train a replacement as well as future legal costs from the EEOC charge.

As always, you will want to speak with the employees when you are making changes in order to soften the impact. People all handle change differently. Some view it as a threat to their current employment. Never focus on the negative concerning changes in leadership but stick to the positive reasons for making the change. It is always a good idea to explain the following:

- What changes are being made.
- How it will impact the individual and the department.
- Why is there a need for the change: what benefits will be gained?

Empowering Your Leaders

It is important that your leaders feel supported and equipped to lead their teams. This is especially true if you are new to your position, or as you make new changes to the plant. You will want to be certain your team stays motivated to succeed. As explained in Chapter 10, there are three types of motivation: Positive, Negative, and Discipline. Positive always works best and almost never requires increases in pay. Develop a plan to promote positive behavior. Some good examples are:

- Lead by example. If you want your people to behave a certain way, show them with your behavior. You set the stage for how your leaders will treat and behave around their employees. Coach them when their actions are not in line with treating people with respect.
- Thank them often for work well done, but be careful to be certain they deserve it. False compliments, even if unintentional, can cause

more harm than good. Others will see the disparity, and will lose faith in you, or view you as ignorant.

- Make sure they are informed and educated on how things should be done. Train them! There are several training programs or manuals you can invest in to help your leadership team understand how to manage employees effectively. You can also invest in books that teach this, and you can hold weekly sessions to train your leaders and supervisors. A once a week lunch with your team discussing questions and answers one chapter at a time will be time and money well spent. As your team gets stronger, let one of your supervisors take the lead role in training.

- Remember to praise in public and reprimand in private. I can't say this enough. While some folks are humble, and don't wish to have the attention of praise (you should honor this), no one likes to have their flaws announced publicly. Constructive criticism is never constructive if done in public, it's just criticism.

- Always be firm, fair, and consistent with every employee according to their function, level of experience, and compensation.

- Don't automatically assume they are incompetent when failures surface. Assess the situation and have a correction plan. Focus on the solution, not necessarily the employee.

Following the Chain of Command

I've always let all of my employees know that they are always welcome to bring issues to me as long as they either have spoken with their supervisor, or it is a situation where they cannot speak with their supervisor, and I always asked them if they had spoken to their supervisor before solving their issue.

Breaking the chain of command can cause confusion at best, but often results in the leader having a more challenging time getting the employees to follow them. It will also consume management time needed for managing costs in the plant, and does not empower the leader to excel in their position. There is no point in having a leadership team if you don't follow the chain of command it represents.

Conclusion

We have now categorized the direct reports in a matrix to be certain they are equipped to handle the position they hold and have determined if any changes need to be made in our leadership team. We have also discussed how to equip the team for success, which includes following a chain of command. In Chapter 12, we will discuss the best way to support our team.

Terms

Chain of command
CNC
Constructive criticism
Direct reports
Equal Employment Opportunity Commission
Organizational char

12

Support the Team

Introduction

Above all else, make sure that all of the employees in your plant understand the plan, and know that you support it. It is a good idea to hold weekly meetings with the employees to educate them on the changes and help them see the benefits.

Ask the employees questions relating to the plan to be sure they understand and help them when necessary. Asking for their advice and/or opinion will help them feel they are part of the solution and will help to ensure they promote it. You may hear excellent ideas, which you might never have thought of previously. If it makes sense, change the plan as long as the goal is the same or better.

Keep a constant check on the progress of the tasks assigned to each leader. Offer help if needed. Be prepared to make adjustments between leaders if needed to keep the plan on track.

Define the Objectives

Everyone in your plant should be aware of the changes and should have had time to voice their opinions on the plan. It is now time to post the objectives of the plan for all employees to see. A good place to do this is on large scale in the break room. This will allow the employees to take their breaks while evaluating the day to day or weekly results. Be sure to keep the information current or take it down. The only thing worse than no information is bad information. If you can get your entire plant working toward the same goal, your job and the work of your leadership team will be much easier, and the results will reflect this. Read Example 12:1 to understand how keeping employees informed on progress translated into cost reductions in workman compensation insurance to over 200k annually.

Example 12:1   Information Strategies

The Problem: The plant was not new, in fact it had been around for over 50 years. The employees were also mostly seasoned veterans performing the same tasks every day, and yet, the plant was failing in almost every aspect. The quality of the product was not acceptable to the main plant where the products were used to complete the finished goods, the efficiencies in the plant were in the 60s, the material variance report showed the plant to be about 20% off mark, safety issues were at an all-time high, and occurred almost weekly, the plant had not been on the monthly incentive for no lost-time-accidents within a thirty-day period in over six months. And to make matters worse, almost the entire group of 60 employees marched into the Human Resource Department to complain about the manager, who was dismissed that same week.

This was the situation I walked into one week after the manager was dismissed.

The Solution: Because I didn't want to end up with the same scenario as my predecessor, I realized that I would have to get the employees in this plant to share the burden of correcting the issues. After getting the detailed performance results for the last twelve months, I put together a meeting format to discuss a few topics on a weekly basis. I was met with quite a few challenges from the supervisors who were sure they could not pull the laminators from their positions to attend a 15–20-minute meeting in the break room every Wednesday. I was relentless, and required all employees to attend, and expected the supervision to find or make a way for this to happen. However, I did allow a select few leaders in Lamination to take turns and either be late or miss an occasional meeting in order to prevent any safety issues from happening in the facility.

The format was consistent every week and started with Safety, followed by Quality, Performance Efficiencies, and Material Usage. The first meeting was an introductory meeting. I asked the employees what we did at the plant, to which they had several answers: parts, boats, lamination, assembly, gelcoat, etc. I then held up a picture of the owner, and instructed the employees that he was the owner, and our function at the plant was to make money for him. That was what we made: Money! If we ceased to make money, there would not be a good reason for the plant to continue. I then let the employees know what our performance as a plant looked like in the areas that we could control to perform the task assigned to us for making the owner money. It was easy to see that we were failing as a team on our responsibilities. And a team we were! I told the team that together we can improve, and it will be a credit to each one of us, or together we will fail, and it will be all of our fault if we do. I let them know that I had one brain, and they had sixty. If they were waiting on me to solve all of the issues, it would probably take sixty times longer than if we worked together. This discussion needed to happen because the employees had teamed together and had separated themselves from the management team, and the plant would not function well until this was corrected.

On the next meeting, I walked into the break room after everyone was inside, and stated the number 9, to which everyone looked at me like I had just come from the planet Mars. I then held up a $5 plant safety buck (fake money used as incentives to buy clothes and other items at the company store) and waved it in the air. Finally, one person shouted, "days safe on the job", and I gave him the $5 safety buck. This was followed by the number of quality issues reported to me from the main plant on completed product, the efficiencies, and the material usage reported by accounting. Brief discussions were had on suggestions for improving these metrics. This was done every week in this format for the entire time I managed this plant. If I was on vacation, one of my supervisors took the lead.

The Result: By the time we had our 4th week's meeting, I had a hard time getting into the meeting area without several employees shouting out the number of days our plant had been without a lost-time-accident. We reached the 465-day mark with no accidents, enjoyed catered lunches weekly, and were told from accounting our focus on safety saved the company about 200k annually in workman's compensation benefits.

Supply the Tools for Success

Once you have defined the objectives, you need to be certain your team has the tools to be successful at attaining the goals. This is not exclusive to tangible tools and equipment, but also includes the ability to perform the work. And, this may also involve the ability to communicate. The stronger your team is, the more chance you have of being successful with your changes. Most importantly concerning communication, if you have a multicultural workforce, it is advisable to learn about the different cultures of your employees. It will help them feel accepted, more at ease to complete their work, and more open to try to communicate solutions. At least half of the workers whom I employed, and who were from other cultures, came to me with college backgrounds. Several were engineers, teachers, and nurses in their countries. This made sense to me when you consider the cost to move from one country to another, especially if the move required air travel. Their degrees were not transferrable, and since they could not speak English, they were forced to take a general labor position. You can imagine how helpful an engineer would be working in a composite facility. All I had to do was figure out a way to teach them to communicate in English. Read Example 12:2 for more information.

Example 12:2   Free English Classes

The Problem: Communication can be a stumbling block in a plant where 75% of the employees were Hispanic, did not speak English well, and were not able to read it at all. A simple task of reading a work instruction to complete a task became a job for two, and one was a lead or supervisor. This was valuable time being spent for every new work instruction submitted.

This really hit home for me when I had to give a positive review to a recent employee. Jaquez was Haitian, he spoke Creole, a deviation of French. He worked for the company for over 25 years and came to the small part plant when he was believed to be too old to produce parts in the large lamination plant. We put him in maintenance as the janitor, a position he excelled at. On one occasion, I brought him to my office to give him his annual review, and he was accompanied by his supervisor. I went through the past few months of his work and told him how happy we were to have him in our plant, and how the work he was doing was helping to keep the plant working. At some point, I realized that although he was nodding his head, he might not be understanding what I was saying, he was just nodding in agreement with my expressions. So, I slowly shook my head yes while telling him that he didn't understand a word I was saying. He shook his head yes and continued smiling. To which my supervisor and I both chuckled. I had given a gentleman a positive review and increased his pay, but he didn't understand a word I was saying the entire time. We later brought up a trustworthy Hattian employee who also spoke English to translate the review again.

The Solution: I called the local college and was told they had English teachers who offered English classes to Spanish adults, but no other language at that time. The teacher would come to our plant two times weekly for 30-minute classes and would teach the employees English. There was a minimal cost, which the company considered an investment, along with the time from their work center. Although this did little to help the Hattian employees who did not speak English, it did help to alleviate the communication issue with all of the Hispanics.

Equipping your employees with the tools to succeed is critical if they are going to have any chance at being successful for helping to achieve your goals.

Expect Results

Once you have outlined the expectation, provided all of the support necessary to be successful, you should expect the desired results from your team. You will want to measure the performance in a group weekly meeting, and you will want to provide a format for the team to follow for this meeting. You should work toward using only fifteen minutes for this meeting, although the first one will probably take a bit longer since your team is getting use to how you use time.

Meetings can be beneficial to keep everyone informed on progress, but they can also become a terrible waste of time if not organized and structured. Read a good example of a fifteen-minute supervisor meeting on the daily schedule in Example 12:3.

Example 12:3   Fifteen-Minute Meeting

While sitting in on a production meeting for a powerboat company in Crystal River, I was impressed to see how the supervision responded to the plant manager concerning their attainment of the daily schedule. There were only a couple of chairs in the office with the exception of the one the plant manager was in, so the supervisor who arrived first sat down, all others stood to give their results for the day. There were about 10 supervisors present.

When the meeting started, it was clear that everyone knew the format for the meeting and followed it. The agenda was to let the plant manager know if you were on or off schedule, and if off, how many hours and people it would take to get back on schedule. Most of the supervisors were on schedule, so their input was fast, just one word. One supervisor was off schedule and requested 4 hours, 4 people on Saturday to get back on schedule. He was granted this, and the meeting ended. To say it was fifteen minutes was probably an overstatement, but by the time everyone entered, gave an update, exited, and exchanged pleasantries, the meeting lasted fifteen minutes.

When you consider the cost of management time and consider how many meetings and for how long the meetings last, it stands to reason that a great deal of cost can be saved if everyone understands the purpose of the meeting and the format for the meeting.

You can always pull individual parties aside to coach them if additional help is needed, but this does not have to be done in a group setting.

Conclusion

Everyone in your plant should stay informed on the progress of the plan, and what goals have been achieved. The employees should understand the details, and feel free to inquire at the area meetings. Adjustments should be made only if required to complete the goal or get it back on track.

Terms

Lost time accidents
Material variance report
Monthly incentive
Multicultural workforce
Safety bucks
Tangible tools
Work instruction
Workman compensation Insurance

References

www.britannica.com/topic/Haitian-Creole

Section 4

Utilizing Free Support

13

Inside Influence

Introduction

Quite often new managers feel they need to have all of the answers for everything that happens in their plant, and are either too proud, insecure, or just don't think to ask others for advice. Or possibly, they don't realize the type of support that is available to them in the form of other managers outside of manufacturing or production processes who work in other areas of the corporation such as Human Resources, Engineering, Procurement, Maintenance, Safety, Quality, Sales, and Accounting.

While these folks may not be as familiar with manufacturing, they can spot things that we might miss. Hence the old saying "can't see the forest for the trees". This chapter will focus on how important it is to utilize free support from people working within your corporation who are not necessarily involved with producing the finished goods.

The Positives from Asking for Advice

Some of the positives gained from asking other managers for their opinions include: Gaining another perspective that might work better than you expected. This person brings with them a different perspective on manufacturing. For instance: You might be watching your employee work and wonder if they are productive and will meet a deadline. Human Resources will probably look at the same employee working and wonder if they are satisfied with their work, their leader, and the department in general. Engineering will look at the same employee and wonder if they are building the product to spec. Procurement will wonder if the supplies are being wasted, and if the system for purchased component parts is working for production. Maintenance is interested if the employee is taking proper care of the tools and equipment. Safety is looking to be certain the employee is wearing the proper PPE (personal protective equipment), and is handling tools and equipment safely. Quality is looking not at the employee, but at the product and wondering if the employee built it at an acceptable level for the customer. Sales is wondering if they can actually sell the product the employee is producing. Production control is making sure there are enough people in the work center for a continuous flow. And, accounting is calculating how much money the employee is generating and/or costing the company, and if the product has a positive ROI.

You can certainly see how each perspective is different, and how you might be able to use some or all of the advice you might be given. A good friend and work companion once told me "I listen to everything everyone tells me, and although I might not agree with what they are saying, I find there is always something that I can use to help me with whatever it is I'm doing". I found these words to be a wise piece of advice, and have learned a great deal of new things because of it.

Another reason to solicit advice from other managers not directly working in production is the ability to help them see what happens on the shop floor, and how their departments affect the ability of the product to be produced. They also gain an additional perspective.

One more thing to note: quite often people do not end up working in the career path chosen from their higher level education. You will read about a CNC salesman in Chapter 15 who had a degree in industrial engineering and helped me on several occasions see a different way to lay out a production process that saved us time in production.

Asking for Advice Shows Your Strength in Character

Asking other managers for their advice also shows that you are humble enough to realize that you don't have all of the answers. Who does? We learn from each other on a daily basis as long as we are on this side of the dirt. Not one person has all the answers because things change too quickly. Only a fool believes they have all the answers. The good news is you don't have to have all the answers, it's just helpful to have several people that you know with experience they are willing to share.

Asking for Advice Helps to Develop Stronger Relationships from Other Areas

Another benefit from asking for help is that you are actually telling others that you value their opinion and view them as a credit to the corporation. You read earlier in Chapter 12 where the plant was instructed that progress would happen much quicker if all 60 brains were working toward the same goal. Consequently, everyone felt empowered, and became part of the solution. If the manager believes they are the only person with all the answers, it is not only going to take much longer to reduce costs in the facility, but it will only have one perspective, and won't be the best plan possible. The following are some of the other areas and how they can and do help manufacturing.

Human Resources

Human Resources typically focus on the how the employee fits into and excels at the position they are employed to fill. They should be an excellent resource for the development and growth of the employee. They will also assist the manager in staying on track concerning labor laws and employee relations and can certainly help to reduce interview time when they learn the management styles of you and your leadership team, and the positions of the employees you manage.

Engineering

Engineering can help the plant by making sure the department has the proper jigs, fixtures, and patterns to accurately manufacture the product. They can also assist with making sure the bills of materials are accurate and the time to produce is measured accurately in the system. Since the cost of goods sold (COGS) is determined in Engineering when pricing the product, it is a great idea to have these folks on your side.

Procurement

The materials group will normally be concerned with making sure the materials to manufacture the product is in usable condition and on hand when needed. They will also be concerned with the amount of inventory in excess of what is needed as it represents money that is not relieved from inventory until the product is complete. They can help the plant by providing station kits for just-in-time work, which will ensure space on the shop floor is available for other processes. They can provide damage reports for manufacturing issues to help identify if additional training is needed. They can source product from other areas to keep production moving. Read Example 13:1 to understand how purchasing came close to shutting down both an Upholstery plant and also the assembly plant it supplied.

Example 13:1   Flat White Vinyl

The Problem: Every morning at 8:30, the managers of all manufacturing departments, as well as purchasing, maintenance, safety, and quality met for a 30-minute meeting to quickly go over issues that would prevent the assembly plant from completing their daily quota.

Normally, the upholstery department did not have any issues as the plant manager worked closely with the buyer to be certain we did not run out of materials needed for the product. But this was not the situation with the flat white vinyl, which was changed to be supplied from a new vendor in China, and was needed in almost every piece of upholstery manufactured. The issue was with the vendor, who instructed the plant that due to a demand issue, they would not be sending our flat white vinyl for another 4 weeks.

The Solution: I instructed the buyer that in another 7 days we would be completely out of flat white vinyl, and three days later, there would be almost no upholstery for the 26 boats we manufactured each day. We needed to find a second vendor, and we needed this to happen quickly. When it became obvious that this was not happening, I contacted one of the former suppliers we had used in the past and had him bring us a few rolls to get by until the Chinese vendor could supply us.

The Result: This did not go over so well with procurement, as they were responsible for raw materials, but it did keep two plants from being shut down.

Maintenance

This is a key department in any plant. Downtime is wasted time and is anti-productive. Preventative maintenance is key to the success of a lamination facility. Maintenance can assist the management on how well the employees take care of the tools and equipment and can help to train on proper equipment care.

Safety

Safety audits will help to be certain the employees focus on safety and work safe. They can assist in chemical training and proper handling of materials and waste.

Quality

The quality department should always have a working relationship with the plant. Quality personnel should be the eyes of the customer. Their goal should be to help the management reduce quality defects on product through training, process, and audits.

Sales

If Quality is the eyes of the customer, Sales are their ears, and can communicate to the plant issues they experience from many angles including quality, fit & finish, engineering, and demand.

Production Control

Production Control is another key department that schedules the work to be started, and the completion schedules for finished goods. In many, composite accounts like products share the same large molds, which presents issues if this department does not understand how build schedules, available space, and equipment issues can affect if the product can actually be started. They also need to be aware of mold availability issues due to excessive mold rework or repair.

Accounting

Accounting is often viewed as the enemy of manufacturing, but actually provide all the necessary documentation to support the actual cost to build the product needed by the management team. They work closely with senior management to set the budget for each area, which can also include CAPEX. It is important to work with this team for the future benefit of the plant, and advisable to find a way to get them to visit the plant on occasion.

Conclusion

It is a great idea to utilize the support and advice of other managers inside the organization who do not work in manufacturing as you will gain perspectives that differ from your own and may provide solutions to issues in your plant that you had not noticed or thought of prior. In Chapter 14, we will look at how Co-Managers within the organization can also help solve issues in your plant.

Terms

Build schedules
CAPEX
COGS
Continuous flow
Demand
Fit & Finish
Fixtures
Jigs
Just-in-time
Mold availability
PPE
Preventative maintenance
ROI
Shop Floor
SPEC

14

Co-Managers

Introduction

Another resource for gaining different perspectives is by utilizing the managers you work within manufacturing. Typically, these are the departments or plants that either provide you with manufactured product to be used in your process, or they are the departments or plants that are considered with your internal customers: the areas that you supply product to, and used to complete the finished product. You will find that both of these areas will provide insight for your process from two different perspectives. The third area that will be beneficial to solicit is the other managers in manufacturing that are not related to the product you manufacture. A mold shop for one of the other plants might be a good example of this as the manager will not be affected by anything that you produce if you are in charge of your own molds.

Your Providers

The departments that provide you with product used in the manufacturing of your finished goods can also offer advice on operations in your plant. It is a good idea to get them to tour your facility and see how their products are used in your finished goods, and what the issues may be with their product. You might be surprised to find out how your internal customers view your product, and how the resolutions can benefit both plants. Read Example 14:1 to understand how an issue with quality completely changed the operation and location of a department resulting in a reduction of costs to the company.

Example 14:1 Sub-Assembly

The Sub-Assembly department was under the responsibility of the mill shop manager, which was one plant located within a large power boat manufacturing site. Sub-Assembly was a different department located in a different area from the Mill Shop plant. They received FRP parts from the Small Part Plant, an offsite FRP plant with a new manager, working for the same corporation; and either installed components on the FRP parts, or installed the FRP parts and purchased parts onto purchased metal frames for the production of hardtop assemblies, console assemblies, seat base assemblies, and hatch assemblies. Sub-Assembly's customer was the Final Assembly Plant for all product at the corporation.

The Problem: The final inspection procedure, which was completed at the end of the production process in the Final Assembly Plant, was written by the Quality Assurance department and worked off by whomever was responsible for the defect. If the defect was part of the FRP product that was installed with the completed Sub-Assembly component, it was the responsibility of the repair technician working for the Small Part offsite plant to correct the issue. On occasion, this meant the part would have to be removed from the sub-assembly and reinstalled, which could potentially take a few hours to complete. Problem #1. There was no formal inspection in the Small Part plant or in the Sub-Assembly plant prior to the final inspection. Problem #2. It was often not possible to determine if the sub-assembly product had been damaged by Sub-Assembly, the Final Assembly plant, or had been shipped defective from the Small Part plant. Problem #3. The Mill Shop manager had a habit of whining loudly to all managers and the boss about the quality of the Small Part product. He was not in the habit of working with his co-workers to correct issues that increased cost in his plant. In his defense, the former manager of Small Parts was not attentive to the issues, and offered little help to correct them.

The Solution: The new Small Part manager had to understand what the true issues were regarding quality with the FRP parts. An inspection process was implemented prior to delivery of the product, and a sign off was required before leaving product with Sub-Assembly. This helped to increase the quality of the small part product, but did not eliminate damage after it was left with the Sub-Assembly department, or determined if it was due to Sub-Assembly or the Final Assembly area. Consequently, the time to repair was still counted against the Small Part plant.

A study was completed by the Small Part manager, which proved that transferring the Sub-Assembly department and responsibilities to the Small Part off site plant (which had ample unused space), would not only solve the issue of damage, since the product would be inspected and a sign off would be required by the Final Assembly plant, but would also reduce the costs of manufacturing incurred by the company due to additional movement. See Figure 14.1.

The Result: The move proved to satisfy the majority of the Quality issues with the FRP small parts, and reduced the amount of time the Small Part repair employee had to spend working on defects after the product left the Small Part plant from 10 hours weekly to 2 on average, a savings of 8 hours for the Small Part plant, which did not include the

Sub-Assembly Location Evaluation			
Located at the main facility - Process	**Labor Used**	**Cost - $35 STD HR**	**Comments**
FRP product received from the Small Parts facility			
Purchased product received and stored at the Mill Shop Plant.	2080 hours annually	$ 72,800.00	7 hrs. daily small parts, 1 hour daily mill shop
Product assembled			
Completed product delivered to Main Plant Assembly stations	3120 hours annually	$ 109,200.00	12 hours daily mill shop
Additional Exceptions requiring FRP work brought to the attentions of the Small Parts facility by the Mill Shop Manager for correction	1040 hours annually	$ 36,400.00	4 hours daily from small parts
Total cost at Mill Shop plant with existing Sub Assembly location		**$ 218,400.00**	
Located in Small parts	**Labor Used**	**Cost - $35 STD HR**	**Comments**
Purchased product received and stored at the Small Part location.			
Product assembled			
Completed Product inspected	1040 hours annually	$ 36,400.00	4 hours daily small parts
Completed product delivered to main plant stations and accepted by the main plant supervision	3120 hours annually	$ 109,200.00	12 hours daily small parts
Exceptions brought to the attention of the Small Parts department for expertise repair as required.	260 hours annually	$ 9,100.00	
Total cost at Small Parts plant for Sub Assembly inclusion		**$ 154,700.00**	
Annual Cost Reduction for moving the Sub Assembly location to the Small Parts facility		**$ 63,700.00**	

FIGURE 14.1
Sub-Assembly Relocation

amount of time in Final Assembly waiting on the repair. The move also reduced the need to move the product until complete and required on the line. The Small Part plant operated on a Just-In-Time (JIT) schedule, so product was delivered daily to stations in the Final Assembly plant. This also helped to reduce damage.

Your In-House Customers

Likewise, it is a good idea to visit the departments or plants that you supply product to, and solicit their advice on how the quality, shipment, and service of your product fits into their operation. I have seen, on occasion, issues that happen through engineering changes which were not communicated correctly to both plants and caused problems in manufacturing. When you solicit the advice and opinions of your in-house customers, you can head off mounting frustrations when production schedules increase, and tempers run short. Read Example 14:2, which describes how simple issues can become bigger problems if left unchecked.

Example 14:2 The Morning Meeting

The Problem: On any given morning, our plant was responsible for providing parts to at least ten different supervisors managing fifteen different lines in the main plant, which produced several types of boats ranging from 16 to 38 feet in length.

Most of the supervisors were on schedule, and this generally meant they would not be looking for any reason why they were behind. However, there were a few that were behind, and of those few, one had a habit of "throwing anyone under the buss" to take the pressure off himself. I did not want to be "under the buss" as had happened in the past with hatches that were believed to be missing from the kit box, and so came in every morning around 6:30 to tour the main plant and ask the supervisor if they indeed had every piece of small part required to get their job completed. Usually I was told yes, but if on the occasion I was not given this answer, it afforded me the opportunity to at least get another part to the line prior to the 8:30 morning meeting. On one occasion we did not have a replacement part for the "missing" item, and it became a hunt and seek mission to find out what happened to the missing part. Missing small parts had become more than a nuisance, and we had to find a solution for this issue.

At the same time, I also checked the boat for a completion status to see if the part that was missing was the only reason why the boat would not complete and found that it wasn't anywhere near being ready as several boxes were still in the kit area and needed to be installed on the boat. This would not solve the issue of the missing part, but it did shed some light on the character of the supervisor.

The Solution: We implemented a procedure to have the supervisor sign off on the kit box that all of the small parts were present and acceptable.

The Result: The main plant still lost, or damaged small parts, but never used the Small Part plant in the morning meeting as the excuse. Consequently, we worked closely with the supervisor to implement a small part return policy to get another replacement part. This allowed us to either repair or scrap product as a write off when damaged, instead of the product ending up in the trash with no accountability. Because lost or damaged parts were being tracked and measured, the frequency of damage was reduced from almost daily to rarely once a month, which represented hours of saved management time and hourly labor to build a new product, as well as lost time in the Final Assembly plant waiting on replacement parts. Both the other supervisor and the Small Part team benefited from this experience.

People make mistakes, we are all human, but it is important for growth that we own up to the mistakes we make. And, if the mistakes of others cause us more work, be careful not to hold a grudge against the other person as this is also anti-productive to what you are trying to accomplish in the plant.

Your Friends and Enemies

Your close friends will give you honest advice if you are open to it, but you may also find that your enemies (or difficult people) you work with may also offer ideas that you had not previously considered. At the very least, by asking the challenging people you work with to audit your process, you will show them that you are open to criticism, and you may just win them over. This does not mean you should take the criticism to heart, but at least you should consider what they have to say. Often the folks that are brutally honest give us the best advice even when it comes across in a negative way.

Conclusion

The benefits from getting advice from other co-managers within manufacturing are more than just the advice they give. It is also a lesson in humility and will go a long way in building relationships with people you work with on a daily basis. Two or more brains working together are usually better than one, and benefit the company as a whole.

Terms

Engineering changes
Finished goods
FRP
Hardtop assemblies
In-house customers
Internal customers
JIT
Kit box
Mill Shop
Seat base assemblies
Sign off

15

Outside Influence

Introduction

Often the biggest impact to reducing costs come to us from areas outside our organizations. This is a big reason why companies hire consultants to help them get organized, eliminate waste, increase throughput and/or quality, improve on safety, train supervisors and leaders, and point out ways to reduce overall cost to the bottom line. In this chapter, we will discover just how beneficial utilizing outside free services can be to reduce overall costs in manufacturing through the use of the existing vendors and distributors we purchase outside products and equipment from; through free management help from retired executives; from attending trade shows; and from other industry managers – non-competitive.

Your Vendors and Distributors

I was fortunate early in my management career to have had a vendor that was always looking for ways to help me be successful. And so, I quickly realized that this was not only a valuable characteristic that I wanted to embrace, but it was also a benefit for the company as a whole, and I began to require this service from all of my vendors. Examples 15:1 through 15:9 are actual issues experienced by composite manufacturers, how the vendor or distributor helped the manufacturer eliminate the issue, and what the reduction in cost equated to.

Example 15:1 Norton – Labor Savings

Introduction to the Problem: The Repair areas in a composite facility often present some of the most aggravating challenges to an area or plant manager. This is mainly due to what would appear to be the inability to consistently complete quality product as scheduled without overtime. I found this especially aggravating when I was managing a small part facility for a large boat manufacture in Florida, and had other management dependent on product from my plant to arrive on time with acceptable quality. The small part plant manufactured and produced 730 parts in 7 stations each day. This represented about 460 different parts ranging from laminated stringers to console assemblies and hardtops. If a key part was not delivered on time, it could shut down the line which produced the end product.

An in-depth repair training program was implemented that encompassed 6 weeks with a senior repair technician. Training focused primarily on chemical processes and usage, tool application, and quality standards. Through measuring outcomes from training, it was determined that a typical repair technician became entry level after 6 weeks of training, completed acceptable work consistently on their own after about 6 months, and became a veteran repair technician after about 6 years.

Training helped, but did not eliminate repair rework due to dull spots, buffing swirls, and stray DA scratches. It was believed by most that these issues were the product of poor gelcoat, ineffective buffing compound, or lower grade sandpaper, until Norton, a brand of Saint-Gobain, and an abrasive manufacturer, recognized the need to develop a training program to help composite customers understand how abrasives worked and how to eliminate these issues by sanding and buffing correctly.

Abrasive grains on sandpaper are fired in the same way pottery is fired in a kiln. This makes the grain very hard, and gives it the ability to cut longer than most technicians will use the abrasive. Abrasive grains fracture when pressure is applied such as through the use of a sander or block. This ideal process is called grain shed, and gives the grain the ability to cut longer as new sharp edges are exposed. A problem arises when the technician does not practice keeping the disc and the area clean during the sanding process. These fired grains if left in the dust too long will attach to the dust and either contaminate the disc (see Figure 15.1a), or will end up contaminating the compound if left uncovered or not wiped from the product prior to buffing.

Often the contamination on the disc presents a defect larger than most grain sizes, which is visible when sanding, and looks like a pig's tail in the dust. This scratch will be impossible to be removed by a buffer after utilizing a normal sanding and buffing grain sequence. See Figure 15.1b, provided by Norton, to see an ideal grain sequence used on production gelcoats.

Abrasive grains in compound are not fired; they are soft, which gives them the ability to break down to the finest grain sizes during the buffing process thus eliminating the finest scratches put into the product by the sanding process. If the technician does not make sure the area they are buffing is cleaned of the fired sanding grains prior to buffing, buffing swirls will be visible. Buffing swirls are a direct result from one of two things: a contaminated buffing pad, or not breaking down the grain in the buffing compound through ineffective use of the buffing tool. This will make sense when you consider the motion of the tool. A simple fix for the contamination issue is to keep the buffing pad and compound clean by covering it when not in use; by cleaning the

FIGURE 15.1A
Disc Contamination

buffing pad prior to covering it, and during the buffing process; and by cleaning the area to be buffed prior to buffing. Do not clean (spur) the pad with any type of wood such as the tongue depressors used in the patch and repair departments. Wood slivers will contaminate the pad in a way that is irreversible.

Often the buffing scratches, dull spots, and stray sanding scratches are not visible directly after buffing because the fillers used in the compounds actually cover up the scratches. But after a few hours or days, these fillers dissipate leaving the scratches visible again.

The Problem: A large manufacturer of sport-boats located in southeast Georgia was having issues in the gelcoat repair areas. Although located in the same city, the company was divided into several different plants depending on boat size and type, and each company used the sandpaper of their choice, without regard to standardization of grain sizes or operator technique. Generally speaking, repair has long been considered 50% process based, and 50% artist based. Basically, the completion of the part or boat was considered dependent on the quality coming from Lamination, and the ability of the repair technician. Training was almost non-existent as gelcoat repair technicians were expected to be up and running within two weeks at this

Finisher Steps

1. Clean the surface
2. Check areas. Only work what is needed. (Work grit sequence backward if necessary)
3. Sand 400 grit, wipe down the area with a non-abrasive cloth.
4. Sand 600 grit, wipe down the area with a non-abrasive cloth.
5. Sand 1000 grit, wipe down the area with a non-abrasive cloth.
6. Buff, wipe down the area with a non-abrasive cloth.
7. Polish

FIGURE 15.1B
Finisher Steps

facility. If the Lamination areas stumbled, additional time would be required due to additional defects coming from the Lamination areas.

Mark Guilfoyle, a manufacturer's representative working for Norton, audited the repair areas in the facility at the request of the VPGM. During the audit, Mark noted that the employees did not use the sandpaper efficiently or effectively to produce the best quality. In fact, due to the way they were using the sandpaper, they were incurring about 25% more time in buffing and sanding to complete the process, and about 10% more in supplies would be required to eliminate the defects the repair technicians were adding to the process. This represented about $420,600 of additional cost spent in labor and materials in the repair areas at the company. See Figure 15.1c to see how this to understand.

The Solution: Training and support was offered to help the repair folks understand how abrasives worked, and how to use them appropriately to reduce or eliminate rework. A grain sequence was given, and the employees were instructed on how to sand and buff on gelcoat. Mark also provided monthly audits of the repair areas and posted these results in areas where the employees could evaluate them. Competition was initiated between the repair areas in order to motivate the

---

The Cost of Repair

Labor:

- **$52,000 per employee per year *(not counting overtime)***
    - ► Based off of a $25.00 standard hour wage. This includes overhead the company incurs on behalf of the employee.
    - $25. x 40 hours x 52 weeks annually = $52,000

- **Doing it correctly the first time will save between 10% - 25% of the labor in this area, which represents a minimum of $5,200 of savings per repair area employee annually.**
    - ► 78 repair employees x 40 hours/wk x $25.00 (Standard hour wage) x 10% = $405,600

Materials

- **Doing it correctly the first time saves about 10% of the abrasives used annually.**
    - ► $150,000 x 10% = $15,000

Total Possible Cost Reduction

- **Labor $405,600 + $15,000 = $420,600**

FIGURE 15.1C
Repair Labor Cost

technicians to follow the process. See Figure 15.1d for an example of the repair audit.

The Result: Because of the strong support from one of the plant managers as well as the senior staff, the repair technicians embraced the changes at this plant, and the rework from the repair areas became almost non-existent. In the Pre-Patch repair area of this plant, the employee head count fell from eight to five employees to complete the daily work, which represented a minimum of about $150,000 annually and included overhead. This reduction in cost represented more money than the cost of all of the abrasives used at the facility in a year. See Figure 15.1e to understand how this was calculated.

Example 15:2 Gelcoat Post-Cure

Courtesy of Tim Reid, Process Engineering and Customer Service Manager

The Problem: A large sport-boat manufacturer on the west coast of Florida was experiencing gelcoat post-cure issues in all of the boats produced with colored gelcoats, but especially in the darker colors. The post-cure would leave the gelcoat looking dull and gave the appearance of "peanut butter" as it was called. Working this issue in the repair area added several hours of labor to each boat affected in the manufacturing process, as well as additional materials for respraying.

Both gelcoat vendors were called in to help the boat manufacturer determine what the root issue was and provide a solution. However,

Norton Quarterly Finishing Audit

Location:

Date:

Use of Abrasive		Results to expect from "No"
Application	Following a recommended sequence? 320/600/1000 or 400/800/1200. Skipping no more than one grain size between steps.	Will spend more time sanding, and more time buffing
Technique	Following a recommended 1/3rd sand over pattern?	More apt to have an uneven surface finish
Productivity	Is the finisher starting with the highest grit size possible to eliminate the defect?	Will spend more time, and risk sanding through the gel.
	Does the finisher remove the previous DA scratches with the subsequent grit size?	The DA scratches will be harder or impossible to remove with the next grain size or buffing compound
Use of Tools-DA		
Application	Starts the DA while it is on the surface and stop the DA while it is off the surface to be worked?	Will spend more time working out deep DA marks.
	Appling the appropriate pressure to the DA to get both directional movements from the tool?	Discs will load up quickly, buff will take longer, sanding will take longer
	Keeping the DA flat while sanding.	
	Is the air pressure correct for the tool?	The DA will wear out faster, and perform poorly long-term. The abrasive will load up faster.
	The DA & Backup pad are in good working condition	Uneven finish results
Technique	Sands in an "S" and then Sideway "S" pattern when possible?	More apt to have an uneven surface finish
Productivity	Sands only the area needed to remove the defect?	Excess labor, materials, and diminished surface quality
Use of Tools-Buffer		
Application	Starts on the part, finishes off.	Risks damaging the gel.
	Keep the pressure around 90 PSI.	The buffer will wear out quicker and perform poorly. It will take longer overall to remover fine scratches.
	Make sure the pad is in good condition free from contamination.	You will take more time, and will have buffing swirls.
Technique	Keep the buffer flat on the part, try to avoid leaning on the edge.	It will take longer, and you risk the chance of discoloring the gel.
	Works all compound off the part in a steady not fast pace.	Does not allow the compound to break down, will not remove the DA scratches.
	Appropriate amount of buffing compound is used	Wasted compound, longer buffing times
Productivity	Able to get sanding scratches out of the part with out putting in buffing swirls.	Wasted compound, longer buffing times
Contamination		
Sandpaper	Cleaning the surface between grit sizes, and when cleaning the disc.	Stray heavy da scratches will be visible in the next grain size.
	Keeping sand paper stored free from contamination when not in use.	Contaminates can cause performance issues leading to longer finish times
	Cleaning the disc periodically to prevent swarf build up.	wasted abrasives, excessive scratches due to buildup
Compounds	Keep paste covered when not in use.	You will take more time, and will have buffing swirls.
	Change brushes (if used) when contaminated.	You will take more time, and will have buffing swirls.
	Clean surfaces prior to buff (wipe down with a micro-fiber cloth).	Buffing swirls
Buffers/pads	Clean buffing pads with a recommended spur or plastic; no wooden sticks.	Buffing swirls
	Keep buffing pad covered when not in use. A shower cap is preferred.	Buffing swirls
Carts	Clean and organized	Less chance of contamination.

FIGURE 15.1D

Repair Audit

neither of the gelcoat vendors were able to solve the issue for this manufacturer.

The Solution: A third gelcoat vendor, who did not supply this boat manufacturer, was also called. This vendor brought in their chemist to evaluate the problem at the manufacturer's site and determined that the humidity in the area was causing the problem. The new vendor developed a promoter package for the gelcoat to combat the humidity, which allowed the gelcoat to cure properly.

Repair Cost Reduction	
8 employees in the department annual cost: 8 x $52,000	$ 416,000.00
5 employees completing the same amount of work after being trained. 5 x $52,000	$ 260,000.00
Total annual Cost Reduction for one department	**$ 156,000.00**
Abrasives used at the facility on average annually	$ 150,000.00

FIGURE 15.1E
Repair Cost Reduction

The Result: The promoter package, provided by the new gelcoat vendor, solved the post-cure problem, and the company saved several hours per colored boat repairing or respraying gelcoat. This represented almost $1,000,000 in annual labor savings alone for respray. See Figure 15.2, for an example of how this was calculated.

Example 15:3 Purchasing EDI

Courtesy of Julie Jenewein, Sourcing Manager American Bath Group, Savannah, TN

American Bath Group is a leading manufacturer of composite bath ware products.

Gelcoat Post-Cure Respray Labor Savings	
Boats affected annually by the Post-Cure Issue	780
Additional Labor hours required to respray each boat affected	32
Total additional labor hours required annually	24,960
Standard hour wage	$35
Cost of additional labor	$873,600
Supplies required to respray	$125,000
Total cost reduction for eliminating this issue.	**$998,600**

FIGURE 15.2
Respray Labor Savings

The Problem: American Bath Group (ABG) was formed over the past few years through mergers and acquisitions thereby combining a variety of cultures, management styles, and legacy systems. This made it challenging to capture the total spend on key commodities.

Maintenance, Repair, and Operations products (MRO) are some of the most challenging commodities to manage for a single company, and required additional time and skill to navigate the complexity of multiple computer systems in order to recognize total spend for each MRO item and the MRO commodity as a whole.

The Solution: American Bath Group's main supplier, Composites One, provided custom reports and Stock Keeping Unit (SKU) rationalization for cost savings through data from their system. This allowed American Bath Group to develop reports that detailed commonality between their plants and companies and helped to identify the alignment of pricing.

The Result: Overall cost savings and SKU rationalization represented a savings of many additional hours spent in managing the MRO supplies. The collaboration between ABG and Composites One captured about $150k from labor, waste, and quality deficiencies.

Example 15:4 In-Mold Coating

Courtesy of Andrew Pokelwaldt, Director Certification, American Composites Manufacturers Association (ACMA), Arlington Virginia

ACMA is the world's largest composites industry association, and supports market expansion, education, advocacy, certifications, composites growth, and overall product marketing

The Problem: A vehicle part manufacturer running a new line of in-mold coated truck parts was having product quality and delivery time issues due to process breakdown and challenges with new in-mold coating process introduction. The issue had cost the manufacturer approximately 70–110k in repair work, labor, spray equipment repair, and wasted chemical gel coat product over the course of one year, which did not include the cost of opportunity lost.

The manufacturer had not developed or followed a proper standard procedure for gelcoat or in-mold coating. They had been engaged by the gelcoat manufacturer but used incorrect spray and equipment procedures to be effective with the gelcoat process. Most of their skilled labor and management team had several years of experience using post-process paint procedures but had limited gel coat experience. They had implemented some production processes that were incorrectly mixing and modifying gelcoat products that would, in many cases, produce porosity-filled products, and in other cases, damage spray tip equipment with spray additives or incorrect initiator levels. As a possible solution to correct the process, the employees made modifications to the process that were approved by the plant engineering management team. However, the technical engineer in charge of coatings had limited or no experience outside of painting procedures.

The Solution: A full understanding of the issue was achieved by the manufacturer after evaluating samples, procedures, and existing problems. The manufacturer was advised by ACMA to bring in the gelcoat manufacturers technical service expert and have them evaluate the issue for a resolution. ACMA worked as an advocate to the manufacturer to help initiate the change. This was a challenge since the process and product were new to the manufacturer and they were hesitant in sharing their process without a non-disclosure agreement. With the technical expertise of the gelcoat manufacturer, the gelcoat application process was modified to help the manufacturer produce acceptable product for their customers. The plant engineer tested the acceptable products, and as the data started to improve, he tracked and reported the results to the senior management.

The Result: The manufacturer experienced improved saving from efficiency gains from cost-effective production on average of 10k each month. This was achieved through savings in labor, material, quality (less rework), and reductions in scrap.

Example 15:5 Tool Crib Time Study

Courtesy of Marshall "Ty" Ellis, VP Business Development, Vallen Distribution, Jacksonville, FL

The Problem: A large cruiser boat manufacturer located in Florida needed to free up necessary space for manufacturing and needed a better solution for dispensing MRO supplies to the employees. The Tool Crib, which housed the MRO supplies and was located at the end of the assembly line, was viewed as an inefficient way to dispense MRO items, and was an obstruction to the assembly flow.

The Solution: Vallen, a large Industrial distribution company, and the integrator for the cruiser boat manufacturer, orchestrated a time study to understand what the actual cost of the current Tool Crib–MRO situation was with the manufacturer, in order to determine if a change in the existing situation would prove a reduction in cost for their customer. Initial evaluation of the time study proved the manufacturer spent on average 7.42 labor hours daily, for a total of 1877 hours annually, which represented a minimum of $51,061 in labor on the current MRO process. This did not include the cost of opportunity lost, which could have been used to produce product. See Figure 15.3 for more details.

Vallen moved the MRO inventory to Point of Use cabinets at the various departments utilizing a Vendor Managed Inventory program to prevent the walk and wait time at the crib window. Vallen was also able to eliminate the tool crib, thereby freeing up the greatly needed production space.

The Result: Vallen was able to help their customer recognize annual productivity savings of $51,061 by moving to a Vendor-managed inventory program at Point of Use. Employees no longer leave their

Tool Crib Time Study	
Labor spent daily for walk and wait time getting supplies	7.42
Available working days	253
Total annual hours spent for walk and wait time getting supplies	1877
Standard Hour cost of labor on walk and wait time	$27
Total annual cost of labor	$51,061

FIGURE 15.3
Tool Crib Time Study

work centers to get supplies, which means an additional 1877 annual hours are available for manufacturing product.

Example 15:6 Lean project/Process Improvement

Courtesy of Marshall "Ty" Ellis, VP Business Development, Vallen Distribution, Jacksonville, FL

The Problem: At a high-volume sport boat manufacturer in Tennessee, weekly orders were turned in by each department at the facility for two different distributors of MRO products. This represented about 40 weekly orders. Purchasing would then have to issue purchase orders for each department to each supplier for that week's requirements. The product was then shipped to the warehouse to be checked in by receiving, and later distributed to the various departments.

The Solution: Vallen, a large industrial distribution company, took over the primary MRO business at the Tennessee plant from the 2 main suppliers, and implemented a new ordering, receiving, and delivering process that eliminated most of the purchasing and warehouse functions related to procuring and distributing the products to the 20+ departments within the plant.

The Result: With the Vallen program, the departments would still create their orders, but it is now easily done through the Internet. Orders come directly to Vallen and are filled and delivered into the plant by Vallen personnel at least twice a week. Purchasing is no longer involved, neither is the warehouse/receiving departments.

In addition, Vallen has worked with the manufacturer's accounting department to streamline the payables process using summary billing, electronic packing slips, and manifests, as well as tracking of the various product accounting codes, all of which significantly reduced the time required to clear invoices for their MRO expenses.

All usage information is reported in great detail on a monthly basis, providing expense data that was previously unattainable, helping each department better understand and control their monthly expenses.

Example 15:7 Accounts Payable

Courtesy of Marshall "Ty" Ellis, VP Business Development, Vallen Distribution, Jacksonville, FL

The Problem: The corporate parent of a high-volume automotive/composite manufacturer changed the invoicing system in place to an Oracle computer invoicing system, which also charges the manufacturer $1.42 per EDI (Electronic Data Interchange) order to process. Due to this change, the manufacturer was processing about 300,000 invoices annually, which created a bottle neck in Accounts Payable.

The Solution: Vallen proposed to implement a Weekly Summary Billing Option, which provides a summary by purchase order number, requisition number, or by charge code. If the "Charge Code Sort Option" is selected, this report groups customer expenditures by cost center for easy identification. Major and minor departments can be indicated, as well as the associate who placed the order.

The Result: Vallen was able to force a transaction to the manufacturer that would be matched, paid accurately, and cleared their AR effectively. Information detail is reported via Access Database on the customer's system, data is fed via FTP transfer of storeroom data off of the Storeroom database server, and the manufacturer can review all data even though it is not on their computer system. This represented a cost savings of approximately $412,339. See Figure 15.4, which illustrates the reduction in cost experienced for this large automotive/composite manufacturer.

Example 15:8 Composite Tooling Shop Syntactic Issue

Courtesy of Alain Lacasse, Owner, NanoCraft Boats, Venice, FL

Accounts Payable		
Issue:		
Charge for each EDI (Electronic Data Interchange)	$	1.42
EDI annual invoices		292,044
Cost of EDI invoicing (1.42 X 292,044)	$	**414,702.48**
Solution provided by Vallen:		
Charge for each EDI (Electronic Data Interchange)	$	1.42
Annual EDI for Weekly Summary Billing Option (reduction of 290,380 EDI)		1,664
New cost of EDI invoicing (1.42 X 1,664)	$	**2,362.88**
Cost reduction for manufacturer through help from Vallen	**$ 412,339.60**	

FIGURE 15.4
Accounts Payable

The Problem: When working in a composite tooling shop, there are only so many material suppliers available who produce machine-able syntactic putties and surfacing primers, and as in this example, it is felt they sometimes take the attitude of "this is what we produce" take it or leave it.

The problem at hand was the plugs created under the 5-axis CNC mill would begin to develop cracks in the finished surface. These cracks would have to be repaired before the plug was waxed for mold construction or take the risk this crack is transferred to the mold surface creating rework to a mold. On several occasions, these cracks would propagate under the geometric nonskid sheets applied to the plug surface, which created a huge amount of rework (80–100 hours).

On other occasions, the plugs were shipped to the customer for mold production but would develop cracks prior to the production process. The customer would conduct the repairs and back-charge the tooling company.

The costs associated with reworking cracking plugs prior to casting the molds were in excess of $90k per year. This did not include the more expensive and harder to calculate costs of losing customer confidence and negative press among customers who experienced these issues.

The Solution: After careful examination of the issue, a decision was made to contact the syntactic supplier to evaluate the problem and advise of a solution. The supplier came in, reviewed the processes, and concluded the cracking did not stem from their syntactic putty, but rather from surface tension created by the primer supplier.

Subsequently, the primer supplier was called in to review the processes, provide an explanation as to why the primer was developing cracks, and to determine why their material was cracking the putty. The following tests were performed:

- Dissecting a plug and examining the cross section of the cracks propagating through the primer and down into the syntactic putty.
- Durometer testing with a Shore D meter to confirm the hardness either was or was not within spec.
- Shrink test of the primer applied to strips of thin cardboard to witness the curl due to shrink.

Their conclusion was that the issue was not their primer, but rather the machine-able syntactic putty was growing globally (in all directions) with temperature changes.

The cracking issues and the finger pointing went on for over a year and were experienced at two well-established and large composite tooling shops. Bottom-line response from the material suppliers was "it's not our material it must be caused by the other guys material" and again, here is our material, like it or leave it. The management team of the composite company felt it was of high importance that a syntactic putty and a surfacing primer were needed from one single source to achieve the best solution possible and eliminate finger pointing between vendors. There would be one phone call to make if a problem surfaced.

Consequently, the friend of a large manufacturer of resins, putties, primers, etc. was contacted to evaluate the issue and give solutions. This company sells product to the other smaller material suppliers who do their "mixing" and then sell it to the public. They do not sell direct to the end user.

The cracking problem and previous evaluations were discussed and shown to the team that arrived from the large chemical manufacturer, and the composite tooling facility was made available to their team for full evaluation and testing for resolution of the issue.

The syntactic putty would be evaluated first, and after experimenting with 5–6 versions of the new material, a putty was determined for the final mix. At first, it was too hard, too much dust, then it went too soft, clogging cutters, hard to sand, etc., but the final version was perfect. This was followed by the new primer, which at first was to sticky, didn't hang well on vertical surfaces, took too long to cure, etc., and after 7 or 8 trials, we had something the shop could work with. Side note: we didn't develop these materials in a materials lab but rather out on the shop floor in conjunction with the employees who work with these materials every day. It was extremely important to have "buy in" from our employees before we approved and moved forward.

The Result: The new manufacturer saw an opportunity and in cooperation with our team, went for it. Their syntactic putty and surfacing primer solved the problem well beyond the normal life cycle of a plug. Plugs and patterns, stored outside for months and exposed to the temperature fluctuations in the Florida weather, did not develop surface cracks.

In one year, the east coast tooling shop would purchase in excess of $700k of materials from the new manufacturer. Rework as a result of cracks became nonexistent, saving hours of labor, schedule impacts, overtime due to rework, consumer confidence, and the reputation of the tooling company to make quality plugs and molds.

Example 15:9 Shop Supply VMI

The Problem: In late 1998, a large high-performance boat manufacturer in Southwest Florida had over 35 areas where shop supply cabinets were stocked and managed. The company had not been able to determine requirements for several shop supplies according to boat produced since many of the supplies were used for several different boats at the same time during the manufacturing and production processes. This included supplies such as spray glue, silicone, disposable cleaning towels, disposable gloves, tongue depressors, razor blades, etc. Shop supply spending was accounted for as a percentage number to the total spend for all products used to manufacture and produce the end product, and was tracked by the purchasing department as a macro number, which made it challenging to measure usage by area, to determine which areas were using more product than required to complete the manufacturing and production processes, and to keep an appropriate level of shop supply stocked in each area.

On occasion, the shop supplies were depleted, and had to be replenished in crisis mode, which caused time constraint issues for the buyers and additional delivery costs. It was believed that an additional three people were required at the facility to control the usage of shop supplies including a buyer, stockroom attendant, and a delivery employee. This did not include the management time at each location to maintain accurate levels, or the downtime costs when shop supplies were not available.

The Solution: AIM Supply provided the company with a vendor managed inventory (VMI) solution. The AIM sales rep would barcode all products for each area, maintain approved levels of inventory at each location, and stock them in the area cabinets on a weekly basis. They would provide usage and cost reports by area and bar code also on a weekly basis, which would allow the area managers to determine and adjust when usage and cost were not in line with production requirements.

The Result: Usage and cost reports were provided to the management team weekly, and were able to be adjusted by the distributor according to the format required with a fifteen-minute turnaround time.

Shop Supply VMI	
Issue:	
Headcount for required employees to manage shop supplies	3
Hours worked annually	2080
Total hours worked to manage shop supplies	6240
Standard Hour Wage	$ 35.00
Cost of Labor for existing headcount to manage shop supplies ($35 X 6240)	**$ 218,400.00**
Solution:	
Headcount to manage shop supplies reduced by AIM VMI	1
Hours worked annually	2080
Total hours worked to manage shop supplies	2080
Standard Hour Wage	$ 35.00
Cost of Labor for existing headcount to manage shop supplies ($35 X 6240)	**$ 72,800.00**
Result:	
Labor cost reduction due to VMI change ($218,400 - $72,800)	**$ 145,600.00**

FIGURE 15.5
Shop Supply VMI Reduction

This allowed for Just-in-Time (JIT) accurate reporting by area, giving the management team the visibility needed for accurate management of supplies as production levels change.

Downtime related to shop supplies became non-existent as levels were adjusted according to production needs at each area.

Two of the required three employees were assigned to other positions which equated to a savings of $104,000. This did not include the additional savings from time previously spent by management at each location to maintain the shop supplies. See Figure 15.5 to see how this was calculated.

Your Equipment Manufacturers

Read Examples 15:10 and 15:11 to understand how your equipment manufacturers can also help to reduce cost in your organization.

Example 15:10 Upholstery CNC

The Problem: The upholstery plant for a large boat manufacturer in south west Florida had a digitized Gerber CNC to cut all materials used to manufacture the upholstery cushions, bolsters, curtains, blankets, and pillows sold with the boats. When purchased in late 1990, the CNC cost around 200k. The company produced 26 boats daily in various colors and sizes. The Upholstery plant cutting room where the CNC was utilized was having difficulty cutting the 15,000 parts daily required to support the 26 boats in a 40-hour work week, and frequently needed to work an additional day on overtime. The cutting room department was operating the CNC 12 hours daily with 9 employees. Upholstery parts were cut by boat first, and by material type second. This meant that the same roll of material would be handled several times during each shift when cutting a different boat.

Also, there was little to no time for preventative maintenance or downtime to replace the CNC table without hurting production. It was apparent a different method or solution would be required to keep up with production without causing additional wear on the machine.

The Solution: The Gerber sales rep was called in to advise on possible ways to gain efficiencies with the CNC without compromising its integrity. The sales rep also held a degree as an Industrial Engineer, and advised for the CNC to cut by material type for all boats at the same time. This meant the upholstery parts would have to be categorized by boat and material type in a marker, and markers for the same material type would be grouped together on a schedule to be cut at the same time. The cut pieces would then have to be labeled and placed in correct bins for each boat for the day's production. Excel was used for the scheduling system. All possibilities for option choices were added to the spreadsheet for selection. See Figure 15.6a as an example of how this information was arranged.

Upholstery Cut Schedule (untouched)

Date	Hull #	Model	Fabric	Gerber Marker	
		23XBR	Flat White Vinyl	113asd	
		23XBR	Flat Blue Vinyl	123sdf	optional
		23XBR	Flat Red Vinyl	123fgh	optional
		23XBR	Flat Green Vinyl	123ghj	optional
		23XBR	.25 Foam Back White	123hjk	
		23XBR	.25 Foam Back Blue	123jkl	
		23XBR	.25 Foam Back Red	123klj	
		23XBR	.25 Foam Back Green	123dfg	
		23XBR	Screen	123kjh	
		23XBRX	Flat White Vinyl	234asd	
		23XBRX	Flat Blue Vinyl	234sdf	optional
		23XBRX	Flat Red Vinyl	234dfg	optional
		23XBRX	Flat Green Vinyl	234fgh	optional
		23XBRX	.25 Foam Back White	234ghj	
		23XBRX	.25 Foam Back Blue	234hjk	
		23XBRX	.25 Foam Back Red	234jkl	
		23XBRX	.25 Foam Back Green	234klj	
		290WA	Flat White Vinyl	456asd	
		290WA	Flat Blue Vinyl	456sdf	
		290WA	Flat Red Vinyl	456dfg	
		290WA	Flat Green Vinyl	456fgh	
		290WA	.25 Foam Back White	456ghj	
		290WA	.25 Foam Back Blue	456hjk	optional
		290WA	.25 Foam Back Red	456jkl	
		290WA	.25 Foam Back Green	456lkj	optional
		290WA	Screen	456kjh	
		290WA	Fabric- plaid	456jhg	
		290WA	Fabric- circles	456gfd	optional
		290WA	Fabric- Mauve	456fds	optional
		290WA	.5 Foam Back White	456dsa	

Upholstery Cut Schedule (product selected)

Date	Hull #	Model	Fabric	Gerber Marker	
1/1/19	DE010	23XBR	Flat White Vinyl	113asd	
1/1/19	DE010	23XBR	Flat Blue Vinyl	123sdf	optional
1/1/19		23XBR	Flat Red Vinyl	123fgh	optional
1/1/19		23XBR	Flat Green Vinyl	123ghj	optional
1/1/19		23XBR	.25 Foam Back White	123hjk	
1/1/19	DE010	23XBR	.25 Foam Back Blue	123jkl	
1/1/19	DE010	23XBR	.25 Foam Back Red	123klj	
1/1/19	DE010	23XBR	.25 Foam Back Green	123dfg	
1/1/19	DE010	23XBR	Screen	123kjh	
1/1/19		23XBRX	Flat White Vinyl	234asd	
1/1/19		23XBRX	Flat Blue Vinyl	234sdf	optional
1/1/19		23XBRX	Flat Red Vinyl	234dfg	optional
1/1/19		23XBRX	Flat Green Vinyl	234fgh	optional
1/1/19		23XBRX	.25 Foam Back White	234ghj	
1/1/19		23XBRX	.25 Foam Back Blue	234hjk	
1/1/19		23XBRX	.25 Foam Back Red	234jkl	
1/1/19		23XBRX	.25 Foam Back Green	234klj	
1/1/19		23XBRX	Screen	234kjh	
1/1/19	HH076	290WA	Flat White Vinyl	456asd	
1/1/19	HH076	290WA	Flat Blue Vinyl	456sdf	
1/1/19	HH076	290WA	Flat Red Vinyl	456dfg	
1/1/19	HH076	290WA	Flat Green Vinyl	456fgh	
1/1/19	HH076	290WA	.25 Foam Back White	456ghj	
1/1/19		290WA	.25 Foam Back Blue	456hjk	optional
1/1/19		290WA	.25 Foam Back Red	456jkl	
1/1/19	HH076	290WA	.25 Foam Back Green	456lkj	optional
1/1/19	HH076	290WA	Screen	456kjh	
1/1/19		290WA	Fabric- plaid	456jhg	
1/1/19	HH076	290WA	Fabric- circles	456gfd	optional
1/1/19		290WA	Fabric- Mauve	456fds	optional
1/1/19	HH076	290WA	.5 Foam Back White	456dsa	

Upholstery Cut Schedule (sorted for printing)

Date	Hull #	Model	Fabric	Gerber Marker	
1/1/19	DE010	23XBR	.25 Foam Back Blue	123jkl	
1/1/19	DE010	23XBR	.25 Foam Back Green	123dfg	
1/1/19	HH076	290WA	.25 Foam Back Green	456lkj	optional
1/1/19	DE010	23XBR	.25 Foam Back Red	123klj	
1/1/19	DE010	23XBR	.25 Foam Back White	123hjk	
1/1/19	HH076	290WA	.25 Foam Back White	456ghj	
1/1/19	HH076	290WA	.5 Foam Back White	456dsa	optional
1/1/19	HH076	290WA	Fabric- circles	456gfd	optional
1/1/19	DE010	23XBR	Flat Blue Vinyl	123sdf	
1/1/19	HH076	290WA	Flat Blue Vinyl	456fgh	
1/1/19	HH076	290WA	Flat Green Vinyl	456fgh	
1/1/19	HH076	290WA	Flat Red Vinyl	456dfg	
1/1/19	DE010	23XBR	Flat White Vinyl	113asd	
1/1/19	HH076	290WA	Flat White Vinyl	456asd	
1/1/19	DE010	23XBR	Screen	123kjh	
1/1/19	HH076	290WA	Screen	456kjh	

FIGURE 15.6A
Upholstery Cut Schedule

At the end of the cutting day, 26 bins (or bags) would be filled with all parts labeled according to part, material type, marker number, and boat number for the sewing department.

The Result: 26 boats, which previously took 9 employees working 6 days (432 hours weekly), and a CNC operating 72 hours weekly, were now able to be completed utilizing three employees working 4–1/2 8-hour days (108 hours weekly), which represents a cost reduction in labor of $842,400 annually. See Figure 15.6b to understand how this was calculated.

These reductions in cost did not account for the additional cost reduction from electricity that was not needed from not using the CNC an additional 1872 hours annually, or costs associated from the wear and tear on the CNC machine, which was now able to be serviced as scheduled on a preventative schedule routine.

Example 15:11 Foam Equipment Maintenance

Courtesy of David L. Travis, General Manager, Composite Research Inc., Blackshear, GA

Composite Research Inc. (CRI) a skiff, flats, and bay boat manufacture located in Blackshear, Georgia since 1994.

The Problem: Flotation foam is required in boats under 20 feet in length per USCG Standards. This foam is a two-part chemical (A/B) that is applied from a special gun. Care must be taken to be certain the foam has the proper ratio of part A and part B, and to prevent the chemicals from curing inside the gun.

The foam equipment would periodically fail due to improper or lack of preventative maintenance, which resulted in foam that did not have the integrity sufficient for a quality product. On occasion, this foam would have to be removed and new foam applied once the equipment had been repaired. The main reason for this issue was the absence of proper training from the foam vendor to use the chemicals and maintain this type of specialty equipment.

The technicians must be properly trained by the equipment or foam manufacturer in order to alleviate integrity issues with the foam, and keep the equipment from failing. The foam vendor that was servicing the account provided quality foam, but was more interested in selling product than helping to train the maintenance technician and foam gun operators on proper start-up, application, shutdown, and maintenance procedures for using their chemicals and equipment.

Downtime associated with this issue equated to about 15 hours of labor per month while the equipment was being repaired, or the ability to manufacture two additional boat per year at $70,000 lost revenue. Labor associated with rework equated to about 10 hours monthly to remove and replace foam for two boats on average that had foam integrity issues resulting from the failed equipment. The annual cost for this issue equated to $73,000, which does not include the cost of the material, additional wear and tear on the equipment due to improper

Gerber CNC Cost Reduction		
Issue:		
Employees required to cut product on the Gerber by product type		9
Weekly hours required per employee		48
Total weeklyhours required		432
Standard hour wage	$	50.00
Total weekly labor cost to complete work	$	21,600.00
Annual labor cost to complete work at 49 weeks	**$**	**1,058,400.00**
Solution:		
Employees required to cut product on the Gerber by product type		3
Weekly hours required per employee		36
Total weeklyhours required		108
Standard hour wage	$	50.00
Total weekly labor cost to complete work	$	5,400.00
Annual labor cost to complete work at 49 weeks	**$**	**264,600.00**
Cost reduction from advice given from vendor		**$ 793,800.00**

FIGURE 15.6B
Gerber CNC Cost Reduction

Foam Gun Cost		
Issue:		
Employees required to remove and replace defective foam due to gun maintenance issue		1
Monthly hours required per employee		10
Total annual hours required		120
Standard hour wage	$	25.00
Total annual labor cost to complete work	$	3,000.00
Lost revenue from boat production associated with this issue (two boats annually)	$	70,000.00
Total cost incurred by this issue	**$**	**73,000.00**

FIGURE 15.7
Foam Gun Cost

maintenance, and additional disposal costs associated with the wasted foam. See Figure 15.7 to understand how this was calculated.

The Solution: After dealing with this issue for over 18 months, we reached out to another manufacture of foam, whose sales rep had a strong technical background, and brought with him the company service technician. Together they evaluated our process, instructed our employees on the proper use of the chemicals and equipment, and trained our maintenance technician on the proper maintenance and repair of the equipment. This company set up a schedule to audit our process every two months, and train any new employees on the use, care, and maintenance of the chemicals and equipment. They worked with the operators to make sure they understood the proper dispensing and shutdown procedures, and provided reports to the plant and lamination managers on the condition of the equipment and any issues from their bi-monthly audits.

The Result: Downtime and rework labor was eliminated due to mechanical issues, which saved the company $73,000.

SCORE and Other Resources like This

Many folks do not realize that there are retired executives who are more than willing and able to offer advice on challenges you may experience managing your plant. These folks often come from backgrounds as CEOs, CFOs, COOs, etc. One group that exists and does this as a free service is SCORE. You can learn more about this organization by visiting their website at www.score.org

They offer free webinars and workshops and will pair you up with a mentor if your request this.

Trade Shows

Another valuable way to gain industry knowledge is by attending composite trade shows such as the Composite and Advanced Materials Expo (CAMX) www.thecamx.org/, the International Boaters Exhibition Conference (IBEX) www.ibexshow.com, and the Society for the Advancement of Material and Process Engineering (SAMPE) www .nasampe.org. At these trade shows, you will have the opportunity to view new technologies in equipment and products, attend conference education sessions, and exchange ideas or concerns with other manu-facturers of composites to help solve issues you might be having. These conferences do cost to attend, but generally pay for themselves with the education you receive.

Similar Industry Managers

Often when we think about sharing information with other similar indus-try managers, we consider trade secrets a protected category that we must keep protected. From visiting hundreds of manufacturers of composites in the last thirteen years, I have learned that there really are no trade secrets in composites that exist outside of the chemicals manufactured specifically for one customer or another. Everyone is basically doing the same opera-tions. If you want your company or plant to have an edge, take care of your best asset, your people.

That being said, welcome other managers into your shop, and take the opportunity to learn from their experience, require the same from them, and offer your advice in return. Some companies already do this, and benefit from not having to learn things others have solutions for. A good example of this was how a manufacturer of Bath tubs and bath ware visited and learned from a boat manufacturer, and attended a training session on gelcoat repair while at the host company. Read Example 15:12.

Example 15:12 ABG at Composite Research, Inc.

A new Quality Process Manager was hired for a manufacturer of bath tubs, showers, and other composite bath ware products, and was interested in learning from other types of composite manufacturing to help reduce costs and gain increases in quality through possible changes in existing processes.

Using the contacts that were available to them, they reached out to one of their suppliers of abrasives, Norton, a brand of Saint-Gobain, to determine if there were other composite manufacturers that might be open to this type of visit.

Always interested in benefiting the industry as a whole, the management team at Composite Research, Inc, which manufactures Sundance boats, opened their doors and allowed ABG to audit their processes for possible improvements to their business. They were also allowed to attend a training session for the repair department regarding Best Practices using Abrasives, a repair training session put together by Norton to help reduce cost and increase gelcoat quality in the repair areas.

The management team and owner, Wally Bell, at Composite Research also met with the team from ABG during the visit in order to exchange information and look for possible solutions to shared issues.

Audits

Another avenue for uncovering ways to reduce cost in manufacturing is by allowing the companies you partner with to audit your processes. Some manufacturers find this a challenge as they do not want competitors to be able to learn their secrets. With the exception of government regulation, from what I have seen in this industry, there aren't too many secrets, and there is a lot of gain from allowing a different set of eyes to uncover things you might not have thought about previously. Read Example 15:13.

Example 15:13 Audit

The purchasing manager, for a large manufacturer of composite columns, came from a different industry in another state, and requested to have an audit performed from the abrasive manufacturer he used at the other location. Norton, a brand of Saint-Gobain, was the abrasive manufacturer, and the composite representative was called in to audit the company processes.

The audit was performed on all areas of the manufacturing as well as maintenance, quality, and shipping departments, and included interviews with the supervision at each location. The manufacturing company officers were extremely helpful and forthcoming with information to assist in the audit. Several cost reduction suggestions were documented and presented to the management team, and a new focus on process improvement and cost reduction was implemented at the company. One item that was suggested was expected to save the company $130,000 annually. See Figure 15.8 for specifics to understand how this was determined.

Plant Audit		
Cost of abrasive discs prior to training on abrasive usage techniques	$	300,000.00
Reduction saving through training and help with sanding techniques		10%
Cost reduction from training	**$**	**30,000.00**
Current disc grain type being used (Aluminum Oxide) annual cost	$	300,000.00
Proposed disc grin type (Zirconia)	$	200,000.00
Cost reduction from grain type change	**$**	**100,000.00**
Total Cost reduction from Abrasive vendor	**$**	**130,000.00**

FIGURE 15.8
Plant Audit

Conclusion

You can see from the several different examples given from vendors and customer that utilized vendors and distributors, that embracing these partners and requiring they help you solve issues related to their products can be very beneficial for reducing cost in your organization. It is also important to be open to new ideas and to listen to advice from others who have either worked in industries similar to yours or have experience managing and are willing to share this knowledge with you.

Terms

Audits
Buffing compound
Buffing pad
Buffing swirls
Charge code
Consultants
Cure
DA scratches
Disc contamination
Dull spots
Durometer
EDI

Fillers
Flotation foam
FTP
Geometric nonskid sheets
Grain sequence
Grain shed
In-mold coating
Initiator
Kiln
Legacy systems
Machine-able Syntactic putties
Markers
MRO
Non-disclosure agreement
Plugs
Point of use cabinets
Porosity
Post-cure
Promoter package
Quality standards
SCORE
Shrink test
Spray additives
Spur
Stringers
Trade shows
Vendor-managed inventory

Section 5

Sustain

16

Measure

Introduction

If you really think about it, we start measuring things almost from the time we are born. Just pass out cookies to three two-year-old children and give one an additional cookie. It will become evident that three is better than two almost immediately. Truth be known, our parents and our doctors are measuring our attainment to standards while we are still in the womb, and for good reason. Measurement is about the ability to make critical decisions at the appropriate time to initiate the most advantageous result that might never have been afforded without measuring an activity throughout the process.

However, there are a few things to consider in order to get the desired results from everyone involved in the measuring process.

- You and your entire team need to understand why there is a need to measure, what the desired results are, and if they are achievable with the fixed resources you currently have?
- How will you measure, and what measuring tools will you use?
- How will you keep your team informed and motivated to achieve the desired results?

Why Measure?

In my travels to numerous facilities, I have noticed a wide spectrum of activity associated with the measurement of everything one can imagine presented in several forms and located in any available space imaginable. I watched as the employees walked past the graphs and charts, and paid little to no attention to them, and wondered how productive the activity to generate these reports proved to be. It was easy to see that some of the reports were old, covered in dust, and had little influence on the activity in the facility. If your reports look like they are old and need attention, your employees will assume they aren't that important as will most visitors in your areas.

Measuring is important. You will uncover opportunities you might not have known were there once you start to measure. You will be more adept to make accurate decisions for positive change, and will be able to see how your decisions will financially impact the results when you measure. But, if you want to get the best results from measuring, you will need to make sure the presentation of the results is in a form that is easily understood by all who are able to view the data, and you will need to be certain the results are communicated to everyone involved on weekly basis to keep your team interested and motivated to achieve the goal. Finally, if you are only measuring to put data points on a chart and do not have a plan to use this information, or have no idea how to use it to help increase financial gains, you are just wasting time.

Read Example 16:1 to understand how measuring helped to increase efficiencies from 47% to 97% in an Upholstery Plant at a large Boat manufacturer in Sarasota, Florida.

Example 16:1 Bill of Material Issue

The Problem: An Upholstery plant at a large boat manufacturer in Sarasota, FL produced all of the upholstered product for the 43 different models of boats. The production completion schedule was set at 26 boats daily. The upholstery plant efficiencies were running on average around 47% from month to month. Material usage on a monthly basis was averaging 100% higher than what was allowed for the upholstery produced. There were several other issues in the plant concerning excessive scrap, damaged product process, warrantee and customer service product integration, employee morale, and a new manager who was not experienced at managing an upholstery plant.

The Solution: The natural instinct regarding efficiencies this low was to assume the employee head count was too high for the product manufactured. The natural instinct concerning waste was to assume the employees were using too much material to complete the product. But both would have been assumptions based on feelings not data. It would be important to measure the current labor hours used to manufacture each individual part that was produced, and compare this to what the Engineering department allowed for each part. The same would be required for the materials used to manufacture each part. To achieve the best results when dealing with labor efficiencies, it is important that most of the employees understand, agree, and work together to achieve the goals. The following plan was implemented to be certain everyone was involved with the process, and were all working toward the same goal.

A spreadsheet was compiled, which detailed each part, and included columns for boat model, part descriptions, part numbers, Engineering labor assigned, believed labor to manufacture, and actual labor used for three consecutive parts. Believed labor to manufacture was achieved by having the crew leaders fill out an Expected Part Labor and Material Usage report for each part. An example of what this basic report looked like can be viewed in Figure 16.1.

What we were trying to accomplish was to determine what was actually needed to manufacture each part concerning the labor and materials. If the Engineering standards were not in line with at most a five percent fluctuation in accuracy, we would have the Engineering department either change the stated labor hours to manufacture the part, or show us how to manufacture the product with the Engineering labor hours and materials stated. Either way, everyone, including Engineering, needed to agree on what the labor and materials required should be for each part. This was critical as the expected costs of goods sold (COGS), profit margin, and consumer price were calculated by using the existing information in Engineering prior to manufacturing the product.

There were roughly 1500 different parts manufactured in the 43 models of boats in the upholstery plant. 75% of these parts followed a process of Cut, Sew, Tack, and Assembly. The other 25% were Cut and Tack only. This meant each operation, Cut, Sew, and Tack, would have to be filled out for each part on the Part, Labor, and Material Usage report. This responsibility was given to the Crew Chiefs, who were responsible for different models in each operation. Eight Crew Chiefs worked at the plant including one for Warrantee & Customer Service parts, one in the cutting operation, three in sewing, and three

Boat Model	23 Fish CC			
Part Description	Port Aft Coaming Pad			
Part Number	ABC123			
Labor Used (min)	Cut		3.00	
	Sew		9.00	
	Tack		15.00	
Total Labor used			**27.00**	
Materials Used	**Material Description**	**Material Part #**	**UOM**	**Quantity**
	Flat White Vinyl	FWV345	SF	0.98
.	.25 Foam back white vinyl	FBW456	SF	6.25
	Welt	WSC01	FT	3.10
	3" HD Foam	3HDF99	BF	0.50
	Silkscreen	SS001	LF	6.00
	Glue	GL002	OZ	4.00

FIGURE 16.1
Part Labor and Material Usage

in tack and assembly. The data was then evaluated by the supervisor for accuracy, and then passed to the plant manager for further evaluation. Exceptions and discrepancies were discussed between the management team until all agreed on the details. Next, actual labor and material usage levels were recorded during the manufacturing process for three of each part built; outliers were accounted for, and an average was achieved to support the plant-stated figures for each part.

Since this information was critical for assigning labor, procuring materials, evaluating labor accurately, balancing work flow, and tracking issues, a completion date of 3 months was set as a goal for all completed parts in the plant. See Figure 16.2 for an example of how this was planned.

The Result: At the completion of the project, all upholstery finished parts had accurate material and labor hours attached to them and either entered or corrected if applicable in the Bills of Materials (BOM). As many as 5% of the parts that were built in upholstery did not previously exist in Engineering. And at least another 40% were missing labor or materials. These might have been due to either a lack of procedure for creating a new part requested from production without Engineering knowledge, or from an absence of Engineering support for new products that were not accounted for through Engineering. Often the upholstery design was changed after the boat was released to production, and the BOM did not reflect what was actually happening in manufacturing. After the changes were in place, the Material usage reports showed the plant with a material waste factor of around 2%. The efficiencies immediately changed from 47% to almost 70% just from correcting the BOMs, which was still a low figure from an efficiency standpoint, but it was a start. This did not represent an immediate cost reduction in labor since the headcount remained the same, but it did allow management to accurately price the upholstery in the boats, and it gave the management accurate visibility concerning the labor used in each operation, which helped highlight where additional attention was needed, and where the next project would focus.

The benefits from measuring the existing situation, evaluating it for accuracy, and making necessary changes to the BOM allowed the management team an accurate way to evaluate labor, load work cells, hold the employees accountable to a realistic efficiency number, and highlight other areas needing improvement.

Measuring Techniques

As mentioned previously, there are many forms of measurement. When selecting how you will compile and present your results, you need to consider your audience, and the goal you are trying to achieve by measuring. Too often the charts and graphs we use are understood well on an 8 ×

Task	01-Sep-01	08-Sep-01	15-Sep-01	22-Sep-01	29-Sep-01	06-Oct-01	13-Oct-01	20-Oct-01	27-Oct-01	03-Nov-01	10-Nov-01	17-Nov-01	24-Nov-01	01-Dec-01
Identify Parts Per Boat Model	Crew Chiefs													
Record time studies for labor and Materials on five of each part		Crew Chiefs												
Assign accurate material and labor requirements for each part								Supervisor & Manager						
Work with Engineering to change the documentation for each respective part per model												Manager & Engineering		

FIGURE 16.2
Material & Labor Accuracy Project Goal

11 sheet of paper in the manager's office, but don't translate to the folks on the floor who don't have the time to decipher multiple layers of information. Be careful to remember your audience when presenting results. Reports, charts, and graphs should be "stand alone" documents: having the ability for the information on the report to be quickly deciphered and understood by anyone who views them on the shop floor. Anyone viewing the information should be able to understand the below points on the information from a quick view:

- What is being measured?
- What the desired result should be?
- What the attainment to the desired result is?
- What expectations or challenges are present that might interfere with the desired result?
- Who is working on or responsible for these challenges?

See Figures 16.3 and 16.4 for good examples of "Stand Alone" reports.

Keeping Your Team Motivated to Attain Results

It is also important to get the involvement of key people in the plant to help support the measurement activities and the documentation that highlights this measurement. Choose employees who are leaders by nature with a positive attitude. Employees who are generally respected by most employees in the plant for their work ethic and ability, and who possess great communication skills. These folks will help to keep the plant report boards up to date, and will assist the leadership team communicating the results to the employees.

The presentation of information in the break and common areas will give your employees additional time to view the data, and bring up any concerns in a neutral atmosphere.

Communication is crucial if you want to keep your employees motivated to achieve the desired results. If you don't communicate weekly on the progress of each project, they will assume it is not important, and won't pay much attention to it on a graph or chart.

Make Sure Your Measuring Stick Is Accurate!

Making sure you are measuring accurately sounds like common sense, but I have found many areas that either don't take the time to make sure what

Carborundum Quarterly Finishing Audit		Robert	Jeremy	Francis	Kerry	Bill	
Location: Plant 3 Repair **Date:** 2/1/19 **Finisher:** Department Totals							
Use of Abrasive							Results to expect from "No"
Application	Is the finisher following a recommended sequence? 320/600/1000 or 400/800/1200. Skipping no more than one grain size between steps.	Yes	Yes	400 600 1000	400 600 1000	400 600 1000	Will spend more time sanding, and more time buffing
Technique	Is the finisher following a recommended 1/3rd sand over pattern?	Yes	N/A	Yes	Yes	Yes	More apt to have an uneven surface finish
Productivity	Is the finisher starting with the highest grit size possible to eliminate the defect?	Yes	Yes	Yes	Yes	Yes	Will spend more time, and risk sanding through the gel.
	Does the finisher remove the previous DA scratches with the subsequent grit size?	Yes	Yes	Yes	Yes	Yes	This will make
Use of Tools-DA							
Application	Does the finisher start the DA while it is on the surface and stop the DA while it is off the surface to be worked?	Yes	Yes	Yes	Yes	Yes	Will spend more time working out deep DA marks.
	Does the finisher apply the appropriate pressure to the DA to get both directional movements from the tool?	Yes	Yes	Yes	Yes	Yes	Discs will load up quickly, buff will take longer, sanding will take longer
	Is the air speed and motion speed within range for the tool?	Yes	Yes	Yes	Yes	Yes	The DA will wear out faster, and perform poorly long-term
Technique	Does the finisher sand in an "S" and then Sideway "S" pattern when possible?	Yes	Yes	Yes	Yes	Yes	More apt to have an uneven surface finish
Productivity	Does the finisher sand only the area needed to remove the defect?	Yes	Yes	Yes	Yes	Yes	Excess labor, materials, and diminished surface quality
Use of Tools-Buffer							
Application	Start on the part, finish off.	Yes	Yes	Yes	Yes	Yes	Risks damaging the gel.
	Keep the pressure around 90 PSI. or 14-2700 RPM	Yes	Yes	Yes	Yes	Yes	The buffer will wear out quicker and perform poorly. It will take longer overall to remover fine scratches.
	Make sure the pad is in good condition free from contamination.	Yes	Yes	Yes	Yes	Yes	You will take more time, and will have buffing swirls.
Technique	Keep the buffer flat on the part, try to avoid leaning on the edge	Yes	Yes	No	Yes	Yes	It will take longer, and you risk the chance of discoloring the gel.
	Appropriate amount of buffing compound is used	Yes	Yes	No	Yes	Yes	Wasted compound, longer buffing times
Productivity	Able to get sanding scratches out of the part with out putting in buffing swirls.	Yes	Yes	Yes	Yes	Yes	
Contamination							
Sandpaper	Cleaning the surface between grit sizes.	Yes	Yes	No	Yes	Yes	Stray heavy da scratches will be visible in the next grain size.
	Keeping sand paper stored free from contamination when not in use.	Yes	Yes	Yes	Yes	Yes	Contaminates can cause performance issues leading to longer finish times
Compounds	Cleaning the disc periodically to prevent swarf build up.	Yes	Yes	Yes	Yes	Yes	wasted abrasives, excessive scratches due to buildup
	Keep paste covered when not in use.	Yes	Yes	Yes	Yes	Yes	You will take more time, and will have buffing swirls.
	Change brushes when contaminated.	Yes	Yes	Yes	Yes	Yes	You will take more time, and will have buffing swirls.
	Clean surfaces prior to buff (wipe down with a micro-fiber cloth).	Yes	Yes	Yes	Yes	Yes	Buffing swirls
Buffers/pads	Clean buffing pads with a recommended spur; no wooden sticks.	Yes	Yes	Yes	Yes	Yes	Buffing swirls
	Keep buffing pad covered when not in use.	Yes	Yes	No	Yes	Yes	Buffing swirls
% Total		100.00	100.00	76.47	100.00	100.00	

FIGURE 16.3
Repair Employee Audit

they are measuring is based off of accurate data; they believe that the data is accurate and won't listen to issues with the data that are highlighted; or they just don't know how to correct the issues that propagate the inaccurate data, and feel that at least the inaccurate data is a starting point at which improvements can be measured against.

Holding employees accountable to a standard that is supported by inaccurate data is counterproductive to what you are trying to achieve when measuring their activity for improvements. This will only frustrate good employees, and cause them to lose faith in your ability to manage the issues. Often the result is a condition known as "learned helplessness", a condition where people who are unable to rectify a situation feel out of control, and eventually accept the condition as it is without the hope of fixing it. They are at a loss of what to do, and therefore do nothing. Read Example 16:2 to understand how this impacts a company.

LAMINATION EFFICIENCIES
WEEK BEG 02/01/19

Decks	Build Quantity	Hull ID	Model	063- Lam
1	1	KA	28SCC	129.473
1		KB	32SCC	0.000
	1	KC	35SCC	153.203
1	2	VJ	3300HCC	240.328
1	1	FL	33C	220.730
2	2	HH	290WA	222.380
		HT	271WA	0.000
1	1	HX	271WA1	109.070
1	1	BG	276WA	93.370
1	1	BJ	225WA	74.980
2	2	BL	245WA	212.162
2	1	HM	250WA	79.758
2	1	VD	2600HWA	68.530
	1	VE	2800HWA	76.500
2	1	VA	2390HCC	99.430
	1	VB	2596HCC	83.920
2	2	VC	2796HCC	168.100
3	3	BF	215DC	209.469
2	3	HS	180DC	126.450
3	3	SF	202HDC	158.880
2	3	HG	210DC	194.040
4	4	VR	190HCC	83.096
5	5	VS	230HCC	131.740
2	2	HJ	22WA	134.460
		HK	22WA1	0.000
4	3	AN	24WA	167.190
3	2	BT	215WA	128.806
3	2	em	*212HWA*	*212.162*
3	2	SC	212HWA	87.500
1	1	ep	*230HWA*	*43.750*
1	1	SD	230HWA	50.650
3	5	HN	180CC	210.750
2	2	eh	*212HCC*	*60.100*
2	2	SB	212HCC	106.280
2	3	BA	175CC	125.784
1	1	BB	200CC	48.657
4	3	BE	201CC	130.971
3	3	BZ	191CC	127.623
3	3	ec	*180HCC*	*159.420*
2	2	SA	180HCC	68.560
1	2	HE	210CC	123.430
2	1	HL	250CC	80.176
5	4	BV	205CC	246.848
1	1	BH	225CC	73.143
	1	BM	245CC	83.658
3	2	ek	*230HCC*	*167.316*
2	2	SE	230HCC	103.300
87	85			5,676.143

		# of Employees	X	# of Hours Worked	=	Reg Hours	Projected Efficiency %
Projected	Regular Hours	113		40		4520.000	107.7%

		# of Employees	X	# of Hours Worked	=	OT Hours	
Projected	Overtime Hours	75		10.000		750.000	

		# of Employees	X	# of Hours Worked	=	Reg Hours	Actual Efficiency %
Actual	Regular Hours	113		40		4520.000	96.7%

		# of Employees	X	# of Hours Worked	=	OT Hours	
Actual	Overtime Hours	75		18.000		1350.000	

FIGURE 16.4
Plant Efficiency Report

Example 16:2 Inaccurate Data

The ability to use data at lightning speeds appeared to take some of the century-old established companies by surprise, and to keep up with the competition, many of them made some of the following crucial errors when setting up their data tracking and report systems:

- No consistency using nomenclature for a systematic identification of products in a company that produces over 20,000 different part numbers. This made it very difficult to group like parts together on an exported spreadsheet for the identification of a product types in one family.
- Exceptions were included into the system that did not respond to the algorithms. Sales reported for specific distributors, where the distributor was attached to one sales rep and the end user was attached to another sales rep, was continually realigned to the distribution sales rep on a quarterly basis leaving both sales reps wondering how the numbers either vanished or appeared on their sales number reports.
- Manual exceptions were allowed as overrides to the information instead of correcting the issues that caused them. These exceptions were often compiled without hard data to support the activity, and were considered a gross error by the sales rep at best.
- Data changes were allowed by various employees working in different departments, and there appeared to be no accountability for why the changes happened.
- Changes to data were not communicated to all parties involved. This made it extremely difficult to have any level of confidence in the ability to accurately be able to report sales data for every sales rep, an issue that not only affected the bonus pay out for the rep, but also reflected on their supposed performance to folks higher in the corporation that were not necessarily aware of the impact the inaccurate data provided.
- The company was not able to utilize the data with any degree of accuracy for regression analysis to help determine where future sales and activity by percentage and area would produce the best results.

Conclusion

When measuring your employees, it is crucial that your measuring stick be accurate, or at the very least you should be communicating with your employees on the actions you are taking to get it to be accurate, and the timeframe involved to make this happen. Do not hold your employees accountable to data points that are not accurate if you have any intention of motivating them positively.

Terms

Balancing work flow
BOM
Crew Chief
Data points
Fixed resources
Learned helplessness
Load work cells
Measuring tools
Profit margin
Regression analysis
Stand-alone documentation
Tack

References

www.verywellmind.com/what-is-learned-helplessness-2795326

17

Accountability

Introduction

As mentioned in the previous chapter, the measurement of activity is at best a waste of time if it is not done to motivate positive change, at least from a human capital position. Accountability is the vehicle used to help drive this change. Every activity performed to generate profit should be measured and compared to a standard, and assigned to an individual for accountability.

Accountability can produce both positive and negative effects depending on how the measurement of the activity is perceived, how it compares to other similar activities measured or not measured, and how the accountability to the measurement is delivered. If the measuring standards are accurate, and everyone is held to the same standard, accountability will help your team recognize the issues, generate the solutions, and own the results for helping your plant reduce operational costs. The old adage of "two heads are better than one" will produce results exponentially when all of your team members are involved, and accountability is consistent and fair.

Make Sure You Are Consistent and Fair with Your Accountability Practices

The goal of accountability should always be a positive outcome. You will want to be certain that you are consistent in your accountability practices. Once you start to measure an activity and are holding your employees accountable to the measurement, you should continue this practice until the desired activity becomes ingrained in the daily habits of all employees responsible for the action.

You should also be consistent in holding employees accountable to the same measurements, or they will consider you unfair, become frustrated, and lack confidence in your ability to manage their actions.

The Positive Effects of Accountability

Accountability does not have to mean reprimand, and if delivered correctly, you should end up with a positive result. Too often we look at the short-term issue or task that did not happen, and forget to look farther ahead when holding people accountable for their actions. Read Example 17:1 to see an unusual way of how accountability was used to gain the necessary support of a supervisor for implementing a process change.

Example 17:1 The Positive Effects of Accountability

The Problem: A facility that supplied 750 manufactured composite small parts to several different lines and stations at a large boat manufacturer had continual issues getting the appropriate parts on time to the designated station.

The facility utilized the same production schedule as the main plant, which only included a hull number and the main plant assembly stations. The small part facility did not deliver the entire boat kit to one station at the same time, but sent pieces as required. The management team at the small part facility had no formal schedule to process this by small part, and since they manufactured the entire small part boat kit at the same time, and well in advance of when it was needed, work in process (WIP) on any given day sat stagnant, was an obstruction to flow, and accounted for about $60,000 USD. This was also an issue for change orders, which affected the large consoles and helms when engine configurations changed the cut-outs for dash panels and gauge components.

Many of the smaller non-changing parts such as hatches were batch built and stacked until needed. Damage to stagnant parts created additional work for the repair area.

The Solution: A detailed schedule was developed to account for the manufacturing and accurate movement of each part required, to eliminate the WIP, and to alleviate the additional work from damage and change orders. The schedule would include labels which were placed on the flanges of the molds prior to gelcoat spray. These labels would be removed prior to trim and placed on the part in an area that would not interfere with production. Larger parts that required additional assemblies would include a work order. See Figures 17.1 and 17.2 to understand what this schedule and labels looked like.

All leads and supervisors at the plant were instructed on the need, benefit, and specifics of the new scheduling system. The small part assembly supervisor was especially anxious to start the new scheduling system and pushed for an early start date. Arranging and printing

Left panel

	Date	Hull Number	Model	Part Description	Do Not Build
x	2/11/19	23IO123	23IO	Bow Hatch	23 in WIP
x	2/11/19	23IO123	23IO	Floor Hatch	
x	2/11/19	23IO123	23IO	Engine Hatch	
x	2/11/19	23IO123	23IO	Port Aft seat Lid	
x	2/11/19	23IO123	23IO	Stb Aft seat Lid	
x	2/11/19	23IO123	23IO	Swim platform	
x	2/11/19	23IO123	23IO	Swim platform lid	
x	2/11/19	23IO123	23IO	Anchor locker hatch	
x	2/11/19	23IO123	23IO	Port helm	
x	2/11/19	23IO123	23IO	Stbd helm	
x	2/11/19	23IO123	23IO	Helm storage door	
x	2/11/19	23IO123	23IO	Stringers	
x	2/11/19	25CR143	25CR	Ski locker bucket	12 In WIP
x	2/11/19	25CR143	25CR	Ski locker hatch	
x	2/11/19	25CR143	25CR	Aft seat storage bucket	
x	2/11/19	25CR143	25CR	Aft seat hatch	
x	2/11/19	25CR143	25CR	Anchor locker hatch	
			19CC	Bait well lid	16 in WIP
			19CC	Live well lid	
			19CC	Bait well bucket	
			19CC	Live well bucket	
			19CC	Console	
			19CC	Port Rod Storage	
			19CC	Stb Rod Storage	
x	2/11/19	32CR012	32CR	Swim platform lid	
x	2/11/19	32CR012	32CR	Stringers	
x	2/11/19	32CR012	32CR	Liner	
x	2/11/19	32CR012	32CR	Cockpit door	2 in WIP
x	2/11/19	32CR012	32CR	Aft storage hatch	
x	2/11/19	32CR012	32CR	Cockpit floor hatch	
x	2/11/19	32CR012	32CR	Rod Storage lid-Port	
x	2/11/19	32CR012	32CR	Rod Storage lid-Stb	1 in WIP
x	2/11/19	32CR012	32CR	Head Door	
x	2/11/19	32CR012	32CR	Head Unit	
x	2/11/19	32CR012	32CR	Galley Sink base	
x	2/11/19	32CR012	32CR	Helm Seat	1 in WIP
x	2/11/19	32CR012	32CR	Helm	
x	2/11/19	32CR012	32CR	Glovebox	
x	2/11/19	32CR012	32CR	Forward storage buckets	1 in WIP
x	2/11/19	32CR012	32CR	Baitwell	
x	2/11/19	32CR012	32CR	Bait well lid	
x	2/11/19	32CR012	32CR	Bait well bucket	
x	2/11/19	32CR012	32CR	Live well lid	
x	2/11/19	32CR012	32CR	Pulpit	5 in WIP
x	2/11/19	32CR012	32CR	Anchor locker hatch	
x	2/11/19	32CR012	32CR	Bow Hatch	
x	2/11/19	32CR012	32CR	Forward floor hatch	
x	2/11/19	32CR012	32CR	Port seat base	
x	2/11/19	32CR012	32CR	Aft Seat Base	2 in WIP

Right panel

	Date	Hull Number	Model	Part Description	Do Not Build
x	2/11/19	23IO123	23IO	Bow Hatch	23 in WIP
x	2/11/19	23IO123	23IO	Floor Hatch	
x	2/11/19	23IO123	23IO	Engine Hatch	
x	2/11/19	23IO123	23IO	Port Aft seat Lid	
x	2/11/19	23IO123	23IO	Stb Aft seat Lid	
x	2/11/19	23IO123	23IO	Swim platform	
x	2/11/19	23IO123	23IO	Swim platform lid	
x	2/11/19	23IO123	23IO	Anchor locker hatch	
x	2/11/19	23IO123	23IO	Port helm	
x	2/11/19	23IO123	23IO	Stbd helm	
x	2/11/19	23IO123	23IO	Helm storage door	
x	2/11/19	23IO123	23IO	Stringers	
x	2/11/19	25CR143	25CR	Ski locker bucket	12 In WIP
x	2/11/19	25CR143	25CR	Ski locker hatch	
x	2/11/19	25CR143	25CR	Aft seat storage bucket	
x	2/11/19	25CR143	25CR	Aft seat hatch	
x	2/11/19	25CR143	25CR	Anchor locker hatch	
x	2/11/19	32CR012	32CR	Swim platform	
x	2/11/19	32CR012	32CR	Swim platform lid	
x	2/11/19	32CR012	32CR	Stringers	
x	2/11/19	32CR012	32CR	Liner	
x	2/11/19	32CR012	32CR	Cockpit door	2 in WIP
x	2/11/19	32CR012	32CR	Aft storage hatch	
x	2/11/19	32CR012	32CR	Cockpit floor hatch	
x	2/11/19	32CR012	32CR	Rod Storage lid-Port	
x	2/11/19	32CR012	32CR	Rod Storage lid-Stb	1 in WIP
x	2/11/19	32CR012	32CR	Head Door	
x	2/11/19	32CR012	32CR	Head Unit	
x	2/11/19	32CR012	32CR	Galley Sink base	1 in WIP
x	2/11/19	32CR012	32CR	Helm Seat	
x	2/11/19	32CR012	32CR	Helm	
x	2/11/19	32CR012	32CR	Glovebox	
x	2/11/19	32CR012	32CR	Forward storage buckets	1 in WIP
x	2/11/19	32CR012	32CR	Baitwell	
x	2/11/19	32CR012	32CR	Bait well lid	
x	2/11/19	32CR012	32CR	Bait well bucket	
x	2/11/19	32CR012	32CR	Live well bucket	
x	2/11/19	32CR012	32CR	Live well lid	5 in WIP
x	2/11/19	32CR012	32CR	Pulpit	
x	2/11/19	32CR012	32CR	Anchor locker hatch	
x	2/11/19	32CR012	32CR	Bow Hatch	
x	2/11/19	32CR012	32CR	Forward floor hatch	
x	2/11/19	32CR012	32CR	Port seat base	
x	2/11/19	32CR012	32CR	Aft Seat Base	2 in WIP

FIGURE 17.1
Small Part Schedule

2/11/2019 23IO123	2/11/2019 23IO123	2/11/2019 23IO123
23IO	23IO	23IO
Stbd helm	Helm storage door	Stringers
2/11/2019 25CR143	2/11/2019 25CR143	2/11/2019 25CR143
25CR 12 In WIP	25CR	25CR
Ski locker bucket	Ski locker hatch	Aft seat storage bucket
2/11/2019 25CR143	2/11/2019 25CR143	2/11/2019 32CR012
25CR	25CR	32CR
Aft seat hatch	Anchor locker hatch	Swim platform
2/11/2019 32CR012	2/11/2019 32CR012	2/11/2019 32CR012
32CR	32CR	32CR
Swim platform lid	Stringers	Liner
2/11/2019 32CR012	2/11/2019 32CR012	2/11/2019 32CR012
32CR	32CR 2 in WIP	32CR
Cockpit door	Aft storage hatch	Cockpit floor hatch

FIGURE 17.2
Labels

the schedule and labels took approximately 15 minutes daily and was required to be completed before the mold lead arrived at 5 am to move molds into place. This responsibility was given to the small part lamination supervisor to produce. The small part supervisor at this facility was very knowledgeable in all small parts manufactured at the plant, and was a key person in regard to the success of the new changes. His input and management of the scheduling system and changes would be required to assure all issues were resolved quickly, and the scheduling system benefited both the small part plant and the main plant. He was not on board with the new scheduling changes, and felt the plant should retain the WIP as additional security when the main plant lost or damaged the small parts, or changed the hull schedule. He was concerned the small part plant would not be able to address these issues quickly, and would cause delays at the main plant. Additional weeks were granted to help the small part lamination supervisor understand the benefits and accept the change, but ultimately a date was set to start the new schedule, which was to be printed on the day prior, and placed on the forklift of the mold lead. When this did not happen as instructed, the plant manager chose to print the schedule and labels, and place them on the mold lead forklift. The plant manager allowed the start to take place without the help of the small part lamination supervisor to make certain the start would happen as planned, and also to show the small part lamination supervisor that the new scheduling system would be implemented with or without his help. A conversation between the plant manager and the small part lamination supervisor to hold the supervisor accountable for not performing would happen later that morning after the successful start of the scheduling system. Accountability would be necessary to ensure the success of the scheduling system, which would be printed daily by the lamination supervisor. Due largely to the time spent making sure everyone understood the need and specifics for the new scheduling system, and the mutual respect between the plant manager and the lamination supervisor, the lamination supervisor initiated the conversation within the first hour of implementation, apologized for not producing the schedule, and completely supported the new process.

The Result: With the help and advice of the lamination supervisor, the scheduling system helped the plant to eliminate the 60k in WIP, eliminated damage from stagnant parts, reduced issues with change orders by at least 95%, freed up necessary space, helped process flow, and helped to hold the main plant accountable for lost or damaged small parts. An additional positive result, although hard to measure, was an increase in trust and respect between the lamination supervisor and the plant manager.

The Negative Effects of No Accountability

Too often I have seen examples where the employees were not held accountable for activity they were responsible for. This is the cause for several issues related to increased cost in manufacturing, including increased labor, quality issues, morale issues, and employee retention. If you fail to hold employees accountable for the activity they are responsible for, they will place no value on the activity. You will also frustrate other employees who try to achieve their best.

Read Example 17:2 to understand how a lack of accountability caused one plant to increase cost over $200,000 by not following management directions.

Example 17:2 The Absence of Management Support

The Problem: A manufacture of boats in Georgia enlisted the help of their abrasive vendor to train their employees on proper sanding and buffing procedures. This was a corporate initiative that encompassed nine separate areas including the component plant.

Each area employees and managers were instructed in a classroom setting on the specifics of how abrasives work, and the recommended techniques and best practices for how to use them. The responsibility of assuring the employees follow the recommended techniques and best practices was left to the management team. The Vendor was responsible for auditing the employees on a continual basis to provide data to the management on progress, motivating the employees to help them follow the recommended best practices by rewarding teams that were successfully following the best practices, and by providing continual training for employees as necessary.

All but one area embraced the changes, reduced their abrasive usage and labor, and reported gains in quality. The management team at the component plant would not hold their employees accountable to the training, and would not support the changes. The vendor refused to audit the employees without the management support as it was viewed as an exercise in futility.

The Result: The employees at the component plant retained control of the quality of the product, the attainment to the schedule, and the amount of abrasives and labor used to manufacture the component parts. Since this plant supplied product to most other areas for this manufacturer, the financial impact to the company due to other areas continually waiting on late product from this plant, the quality issues

The Cost of No Accountability		
Expected Cost		
Materials	6500 boats annually to be repaired in acceptable conditions	$48,000
Labor	25 employees at $25.00 Std Hr Wage	$1,300,000
Total Expected Cost		**$1,348,000**
Actual Cost		
Quality Issues after Completion	8 additional hours daily spent reworking unacceptable product at other plants.	$52,000
Abrasives used	.04 more abrasives required for loss of savings and quality issues	$49,920
Labor used	25 employees at $25.00 Std Hr Wage + 25 Hours overtime weekly (premium not included)	$1,332,500
Total Actual Cost		**$1,434,420**
Difference between Expected and Actual		($86,420)
Proposed Cost		
Materials	6500 boats annually to be repaired in acceptable conditions	$48,000
	Reduction of 10% Materials from process improvements	($4,800)
Labor	25 employees at $25.00 Std Hr Wage	$1,300,000
	Reduction of 10% Labor from process improvements	($130,000)
Total Proposed Cost		**$1,213,200**
Difference between Proposed and Expected		$134,800
Difference between Proposed and Actual	**Cost Reduction not recognized due to no accountability**	**$221,220**

FIGURE 17.3
The Cost of No Accountability

that had to be addressed, and the increase in materials and labor to repair the product ranged more than $200,000 annually. See Figure 17.3 to understand how this was calculated.

Conclusion

Accountability will help you achieve your goals if your employees understand what is expected of them, if they have the ability to perform the task, and if the accountability is delivered professionally to them. The absence of accountability will have a negative impact on your plant.

Terms

Change orders
Exercise in futility
Flanges
Gauge components
Hull schedule
Process flow
WIP

References

www.inc.com/partners-in-leadership/how-positive-accountability-can-make-employees-happier-at-work.html
www.biblegateway.com/passage/?search=Ecclesiastes+4%3A9-12&version=NIV

18

Reward

Introduction

When we think about reward, we generally think about some type of financial gain for performing an outstanding task, attaining a milestone, or achievement of a difficult goal. Reward is the prize we get for positive results above expectations, and should be used in business to promote additional positive behavior and results that reduce cost.

There are several types of reward that can be used to motivate employees to perform above expectations. Some of these rewards will motivate short term such as a close parking spot for the month for being a safe employee, and some have lasting effects for life such as public praise for a job well done. Some reward systems are long term, and require continual attention such as a bonus program for exceeding required expectations.

It is important to understand what motivates the person you are trying to recognize when giving a reward since reward can mean different things to different people, and the effect you are trying to achieve might not be what you end up with.

Types of Reward

To most people, reward means financial gain. After all that is why we work, correct? There are several types of financial reward including the following:

- Performance Bonus is generally based off of gains above an expected goal, and is usually paid in whole or on a percentage of these gains. It should be quantifiable, easy to understand and calculate, and fair. In the article How to Structure an Employee Bonus Plan by Anastasia in Cleverism.com, Relevance is mentioned as one of five elements that are needed for the plan to be effective. Anastasia states "It should be meaningful to everyone – the management and the employees. There should be meaning attached to the bonus, so that the employee will feel a higher sense of fulfillment, knowing they are receiving the bonuses because they deserved it."

When a bonus plan is effective, your employees will actually help you manage the activities and assets of the business. Read Example 18:1 to learn how a performance bonus plan benefited a company in Pipestone, MN.

Example 18:1 Performance Bonus

While working for a large manufacture of boats with over 25 locations, I was sent to a sister location to learn how Inboard/Outboard (I/O) motors were installed, as this was an activity expected to be performed at the plant where I was employed. A performance bonus program was utilized at all locations, although the plant where I worked was relatively new, and had not yet embraced the plan in a way that was beneficial to both the company and the employees. This was largely due to poor communication of the performance plan.

Having been assigned to the Engine Installation area at the sister location, I was able to learn from two employees worked in stations side by side installing I/O motors. The two stations were designed to be identical. All hand tools for each station were located on easily accessible boards or airlines, which made locating and using them almost effortless. The employees worked at their tasks in an almost robotic flow. On one occasion, one of the employees discovered a better way to perform a task, and shared it with the other employee. They then quickly agreed on who would let the night shift crew know to change the process to the new way for this task, and who would let the engineering team know to evaluate and change the Build Book.

The work ethic of these employees was impressive, and I wondered if they actually worked like this every day or if some of this was for show since they were training a salary employee. The answer came after another week of evaluation and training. A relatively new employee had almost reached his 30-day temporary employment deadline, and all of the employees in the assembly plant were meeting to decide if the new hire should be retained or let go. It was a simple yes/ no vote, and the new employee was retained as a permanent employee. I was new in management, and had never seen or heard of the hourly employees having the ability to decide the employment status of another hourly employee.

After questioning the management team, I was told this was a direct reflection of the bonus plan the hourly employees benefited from. The assembly department earned a performance bonus as a team. They were able to earn a percentage of profit above the standard hour cost of manufacturing the boat, and this money was divided between each employee in the department according to how many hours they worked in the month. This bonus was also almost as much as the sum of their monthly paychecks, which is why they were also concerned with how they worked, and who they worked with. Figure 18.1 is an example of how this plan was calculated.

LAMINATION EFFICIENCIES

Month BEG 02/01/19

Build Quantity	Hull ID	Model	063- Lam
4	KA	28SCC	517.89
4	KC	35SCC	612.81
8	VJ	3300HCC	961.31
4	FL	33C	882.92
8	HH	290WA	889.52
4	HX	271WA1	436.28
4	BG	276WA	373.48
4	BJ	225WA	299.92
8	BL	245WA	848.65
4	HM	250WA	319.03
4	VD	2600HWA	274.12
4	VE	2800HWA	306.00
4	VA	2390HCC	397.72
4	VB	2596HCC	335.68
8	VC	2796HCC	672.40
12	BF	215DC	837.88
12	HS	180DC	505.80
12	SF	202HDC	635.52
12	HG	210DC	776.16
16	VR	190HCC	332.38
20	VS	230HCC	526.96
8	HJ	22WA	537.84
12	AN	24WA	668.76
8	BT	215WA	515.22
8	em	*212HWA*	*848.65*
8	SC	212HWA	350.00
4	ep	*230HWA*	*175.00*
4	SD	230HWA	202.60
20	HN	180CC	843.00
8	eh	*212HCC*	*240.40*
8	SB	212HCC	425.12
12	BA	175CC	503.14
4	BB	200CC	194.63
12	BE	201CC	523.88
12	BZ	191CC	510.49
12	ec	*180HCC*	*637.68*
8	SA	180HCC	274.24
8	HE	210CC	493.72
4	HL	250CC	320.70
16	BV	205CC	987.39
4	BH	225CC	292.57
4	BM	245CC	334.63
8	ek	*230HCC*	*669.26*
8	SE	230HCC	413.20
340			**22704.57**

Cost of Labor allowed	22704.572	X $35	$794,660.02
Cost of Labor used	19765.000	X $35	$691,775.00
Difference			$102,885.02
20% To be divided up between the employees.			$20,577.00
Extra Pay Per Hour Bonus	$0.96		$1.04
Average bonus per month per employee	128ee		$166.57

FIGURE 18.1
Performance Bonus

- Material Savings Bonus is given for using less shop supply materials to manufacture product than is expected. It is often difficult in a multi-faceted operation, such as manufacture of boats, cars, or airplanes, to set a finite standard for how much shop supplies are required per product, area, or employee. This is largely due to variability in the chemical processes. For instance, if a laminated product has additional defects due to chemical integrity or weather changes, more shop supplies will be required to complete the process. Implementing a material savings bonus is a great way to get the employees to think about materials as if they purchased them, and assure the percentage used is as accurate as possible. You should gradually adjust the percentage for shop supplies as the employees help get the levels more accurate, which will affect the checks they receive. As new product is developed or changes are made to existing product, they will experience fluctuations in shop supply levels. Always give them something to achieve if you want them to continually think about shop supply waste. Read Example 18:2 to see of how a small composite manufacture was able to save money with shop supplies by implementing a Material Savings Bonus program.

Example 18:2 Material Savings Bonus

It was strongly believed that the shop supplies were being abused and used excessively in a high-speed composite facility in North Florida. This thinking was supported by evaluating the garbage cans at the end of each shift. Many good items were found in the garbage, including stainless steel screws, nuts, washers, and other fasteners such as cable ties and clamps; caulking tubes that still had material in the end of the tube; sandpaper rolls that were not completely used; many disposable gloves, face masks, and aprons; and on occasion, small perishable tools such as retractable knives, putty knives, drill bits, tongue depressors, and micro-fiber cloths.

It was clear the employees would need to be informed of the costs of these items, and instructed on procedures for disposing of them. Management felt certain a Material Bonus Program could help identify correct levels of shop supplies, and get the employees to think of the shop supplies as something they paid for themselves. It was believed that the company could reduce shop supply cost by $152,327 annually, which would increase profit by $76,163. See Figure 18.2 to understand how this was calculated.

A simple plan was developed which took monthly shop supply purchases from the past few years and compared that to the amount of finished product and number of employees to get an average amount of shop supplies by product produced required. See Figure 18.3 for an example of how this was calculated.

Shop Supply Waste	Column1
3 year average of COGM	
Year 1	$ 77,060,654
Year 2	$ 81,500,923
Year 3	$ 75,786,956
Average	$ 78,116,178
Shop Supplies are .03% of COGM (as taken from a three year average spend)	$ 2,343,485
Reduction estimate based on evaluation of wasted supplies at .065%	$ 152,327
50% split for the company (annual)	$ 76,163
50% split for the employees (annual)	$ 76,163
Expected quarterly gain for the 103 employees ($78, 163 ÷ 4)	$ 19,040.82
Expected quarterly bonus for each employee ($19,041 ÷ 103)	$ 185

FIGURE 18.2
Shop Supply Waste

Supply Material Requirements	Column1
3 year average of COGM	$ 78,116,178
3 year average Shop Supply Purchases	
Year 1	$ 2,561,068
Year 2	$ 2,280,496
Year 3	$ 2,188,892
Average	$ 2,343,485
Shop Supplies as a percentage of COGM	3.00%
Average annual Finished product units	$ 3,120
Average shop supply per unit ($2,366,637 supplies ÷ 3120 Units)	$ 751

FIGURE 18.3
Supply Material Requirements

Once the average amount of shop supplies by product type was determined, it would be compared to what was currently being used and product produced on a quarterly basis, and a percentage of the cost reduction would be divided up among the employees based off of regular (non-overtime) hours worked. The plan was put into action, but getting the employees to look for ways to not waste the supplies took six months and two bonus checks. The first quarter produced checks for the employees at just under $100. The second check reached to around $125, and by the third quarter, the Material Savings Bonus checks were running around $160 per employee. This equated to about $0.31 per hour per employee on a quarterly basis, which was a nice improvement for both the company, and the employees.

On one occasion, a new hire was sweeping the floor, and swept up a disc roll of sandpaper. He was about to dispose of the disc roll with the other floor garbage when another employee explained that the disc roll he was throwing away was part of the employee Material Savings Bonus. This was an excellent example of how the employees will help manage waste when it affects their ability to earn additional income.

- Tuition Reimbursement is a win-win for the employee and the company as long as the employee stays with the company, although it remains a benefit for the employee regardless. The benefits from a tuition reimbursement program include the monetary reward, gained knowledge, and often a more productive employee. In an article by Maria Coppola titled Tuition Reimbursement at Work is a Bonus, Coppola states "Continuing education, especially if it results in a degree or certification, is equal to getting a raise at work – it puts dollars in your pocket and represents a life-long achievement." This benefit is one that often gets overlooked by employees. Good management will encourage their employees to take advantage of this bonus as the more educated your employees are, the more they feel empowered to perform, and the easier your job becomes.
- Non-financial rewards (at least from the perspective of the employee) are often more valuable to the employee than financial rewards, and if delivered correctly, their effects and benefits can last much longer. These rewards can be as simple as getting a free vacation day for outstanding work, or for giving the company an idea that resulted in a process improvement, which reduced cost or increased throughput for the company.

Another reward that will not only reduce cost associated with work compensation and accidents, but may also increase quality, is safety incentives like holding catered lunches for milestones reached with no lost-time-accidents. This reward, if provided correctly as we read in Example 9:2, will save work compensation insurance costs, possible lawsuits, and personal injury.

Company-sponsored events are also great ways to reward employees and let them know management cares and appreciates them. Companies typically enjoy holiday lunches for their employees, but it might be a great idea to consider a separate Employee Appreciation Lunch during a time when there is no holiday. Vendors can donate items for raffles, and/or as giveaways. The idea is to appreciate the employees, and make sure they know they are important to the company. Recognition and praise for work well done is always a reward that is remembered long after the employee has retired. But, make sure the employee is deserving of the praise, or their co-workers will be demotivated by watching an underserving employee get accolades, and as negativity breeds negativity, you will lose credibility with your team if you reward without good reason.

Understand Which Rewards Motivate Your Employees

You need your employees to become engaged with the company, and take pride in the work they do. You also benefit from employing them when they feel empowered to help make decisions that affect the financial outcome of the company in a positive way. If too much time spans between rewards, your employees will lose interest and work will become mundane to most of them. Saying "Thank you" and recognizing on a continual basis will help your employees strive to perform their best. Performance bonuses or at least the fair tracking of an annual bonus will help them stay motivated to press on toward the goals they have been given. Woolley and Fishbach write in their article titled: It's About Time: Earlier Rewards Increase Intrinsic Motivation, "More evidence suggests immediate rewards are beneficial, ... They're a useful tool for increasing interest in an activity." These immediate rewards can be monetary or non-monetary.

When Things Go Right

Don't forget to take full advantage of highlighting the times when positive results show outstanding performance. A good example of how bringing recognition to one lamination deck crew helped the entire lamination deck crew department reduce air voids in the laminated product from 30 inches on average to almost zero in a matter of a few weeks is given in Example 18:3.

Example 18:3 Lamination Deck Air

A High-Performance Boat manufacturer had four deck lamination crews, which consisted of one chopper gun operator, and four lamination deck rollers. All four deck crews were responsible for manufacturing all decks at the facility. All four crews were also responsible for leaving, on average, 30 inches of air voids from not rolling the fiberglass out correctly. This was not acceptable. After evaluating all of the deck crew employees and their supervisor, it was determined that attention to the work they were doing would be the first corrective measure utilized. The management team and quality assurance would post on a 4-foot by 8-foot dry marker board located in the lamination department which listed the decks manufactured, the name of the crew chief, and the amount of air voids left in the completed deck. This information was visible where all employees could see. A total amount of air for the week was also recorded.

As each completed deck was recorded, the employees began to pay attention to the board, and natural competition began. It was clear they were interested in which team had left the least amount of air voids, which meant they were the best lamination deck crew.

At the end of the week, the manager of the lamination facility took a box of 100,000 dollar chocolate bars, and stood on a crate in the middle of the plant floor, and called the deck crews to a meeting.

> Deck crew number 3 is the best lamination deck crew we have in this facility, she stated, and I want to thank you for your efforts, and give you these $100,000 dollar candy bars as a gift for building the best decks this week.

What happened next was expected because the lamination manager understood her employees, and what motivated them most to perform above the performance standards. The following weeks the deck crews checked the board frequently to see where they were in comparison to the other crews, and at the end of the next week, a different crew was on top with the least amount of air voids. They went to the manager's office requesting their reward, which wasn't the candy bars, it was the recognition in front of the rest of the plant for being the best. After a few weeks, the air voids were down to between zero and three voids. This equated to less time in repair, and a better finish for the customer. Reward is not always a tangible item.

When Things Go Wrong

Derek Stockley, a training and performance management consultant, states in his article *Achieving Fair Financial and non-Financial Rewards* "Employees want the rewards to be shared fairly and equitably. If they are not, dissatisfaction can cause severe morale and performance problems". I find this statement to be 100% accurate, and surprisingly people will actually leave a very well-paid position for a lesser position with another company that treats them fairly. This is not a reward issue, it is a trust issue.

When the employee views the reward as unfair, they will have a hard time with trust concerning their manager or employer. Once trust is gone, it will be difficult to get it back, and almost impossible to motivate this employee to perform above standard until they trust you. If there is a misunderstanding regarding reward, you should help the employee understand why the reward is fair.

Conclusion

Rewarding your employees will motivate them to perform above standard as long as you understand what motivates them, and their perception of the reward is fair. Often the reward that motivates them the most does not add cost to the company such as recognition and praise for work well done. These are simple rewards that have long-term effects because the employee feels appreciated.

Terms

Air voids
Bonus program
Build book
Chemical integrity
High-Performance Boat
I/O Motors
Material Savings bonus
Milestone
Non-financial rewards
Performance bonus
Standard hour
Tuition reimbursement

References

www.derekstockley.com.au/a-financial-rewards.html
https://corporatefinanceinstitute.com/resources/knowledge/accounting/product-costs/
Coppola, M. Tuition reimbursement at work is a bonus. *Expertistas.com*, July 2016.
http://expertistas.com/2014/12/30/tuition-reimbursement-at-work-is-akin-to-getting-a-raise/aboutme
Woolley, K., Fishbach, A. It's about time: Earlier rewards increase intrinsic motivation. *Journal of Personality and Social Psychology*, 2018; 114 (6): 877. DOI: 10.1037/pspa0000116. www.sciencedaily.com/releases/2018/06/180606143709.htm

Additional Reading

www.rrgexec.com/rewarding-your-employees-15-examples-of-successful-incentives-in-the-corporate-world/
www.cleverism.com/structure-employee-bonus-plan/

19

Change Management

Introduction

Change is not always viewed in a positive way for many managers. Consistency and organization are easier to manage than the outcome uncertainty and employee obstinance that change often brings. But change is required in order to be competitive when new technologies arise, or if you want to take the lead by becoming an innovator through forward-thinking. Managing down when production levels change can be especially difficult as it requires reallocating or eliminating positions, but the absence of making these necessary changes not only have a negative effect on the balance sheet, they hinder the employees from working in the "zone", which increases management time, safety incidents, and quality defects.

Training your replacement is another challenge for many managers, although it is crucial to keeping the plant operating efficiently, effectively, and allows for better decision-making. Two heads are often better than one especially when they work toward the same goal.

Managing Down

I first heard the term, Managing Down, and thought it an odd way to describe a lay-off. When there are too many workers to perform a function, it is crucial to the business to remove the ones that are not required. This does not have to mean the employee(s) no longer work for the company, but it does mean the unnecessary worker(s) should be assigned to another position if possible until needed for that function again.

When production schedules change, and less product is being manufactured, employees will naturally slow down, often this happens without thinking about it. It is quite normal to take a sigh when the workload is lighter. Much the opposite as when we are faced with several reports that have to be processed. Most of us generally work at our best when gently pushed by deadlines and unfinished product. In many types of processes, this is called working in the zone. Basically, the subconscious takes over most of the repetitive work, allowing the rest of our resources to handle the exceptions well. Work is completed with higher quality, less safety issues, and higher efficiencies. I have found that in manufacturing, and as long as you provide your employees with the necessary tools and knowledge to perform the task, all that is needed to get them to work in the zone is a sense of well-being with their job, and a gentle, constant push of work that they can achieve. Most people naturally feel a sense of accomplishment for being able to do their work well. When this is reinforced with appreciation from their boss, it will generally continue.

If, however, the work is not available, it will be impossible to work in the zone as you need the rhythm of constant work to push you. When there are too many employees and not enough work, the entire department they are working in will suffer. This is especially true for the repair areas in a composite facility, which is challenging to adhere a consistent completion time limit when work fluctuates due to lamination defects and operator skill, and if well-defined quality standards for gelcoat are not established. It requires the supervision and management to pay closer attention on a continual basis regarding work load in order to adjust before repair workers create more work (some intentionally, some unintentionally). Make sure you explain the reason for the change to your employees at all levels. Helping them understand will alleviate future issues resulting from gossip and negativity. Read Example 19:1 to understand how this negatively affects the repair area and the plant.

Example 19:1 Repair Control

A large composite facility in Orlando, FL was having issues in their repair areas with gelcoat finish quality and schedule attainment.

Warrantee for this company recognized gelcoat defects as 76% of their claims turned in. Because the completed product was too large to send back to the plant for repair, the customer was allowed to hire a sub-contractor specializing in gelcoat repair to correct the issues and bill the company. This cost was over 10 times what it would cost the plant to correct, but still less than the cost of sending a company repair technician to complete. Consequently, it was a challenge for the company to determine if the cost of sub-contract repair was fair or necessary.

Additionally, since other areas of the plant depended on the repair areas to complete and deliver product to the next operation according to the schedule, when the repairs were not completed, the subsequent operations suffered, workers had to be reassigned to other areas, and a bottleneck in the repair area was evident.

To make matters worse, the repair department, which was composed largely of immigrants from Vietnam, behaved as a team and followed the queue of their self-appointed leader, the only one in their team who understood and spoke English. It was thought by the company that if pressed too hard for change, the entire "team" might quit.

The plant engineer reached out to their vendors for additional help to determine if proper sanding and buffing practices were utilized, and to see if a different type of abrasive might help with throughput. An audit was performed, which proved that the majority of the repair technicians did not understand how to sand properly, and did not follow a proper grain size sequence. This was the underlying reason for at least 50% of the warrantee issues.

A blind test was scheduled for the following week, which encompassed utilizing one repair technician to test three different products with two different grain sequences. Followed by another repair technician to buff all three areas as one area. The change in technician was done in order to be certain the technician did not intentionally fail the test by not using the tool the same on each area. Profilometer readings would be taken after each change in grain size and type, gloss readings would be taken after buffing was completed, time would be recorded for each step, and all materials would be accounted for.

The sanding part of the test went as planned. Five-inch Dual Action (DA) pneumatic sanders were used with the corresponding discs. Correct sanding technique is easy to determine by placing two marks on the edge of the DA backup pad, and watching for a strobe effect while sanding. If the technician is pressing too hard, the marks will hold to their position, and sanding will produce deeper scratches. Dual action happens when the backup pad is orbiting and circulating at the same time. See Figure 19.1 for an example of what these actions look like.

After sanding, the first repair technician attempted to buff the areas, and was told to go to the supervisor's office. He immediately took offense to this and wanted to finish the test. He was allowed to return to his work station before the test was complete, which was located next to the area where the test was performed, and managed to roll his Mechanic's Creeper over to the bottom side of the test product where he called to the second technician in Vietnamese, and started giving instructions. It was believed that he was trying to let the technician know which area to buff correctly on, and which ones to fail. He was chased away from the test, but it wasn't possible to understand if the instructions he gave actually hurt the outcome.

It would have been more beneficial if the management team had taken the employees aside and explained the need for the test, and what the company's intention was for change. Often employees think the company makes changes with product to lower the cost, and look for cheaper products to accomplish this, which make their work harder to perform. And I have also seen repair technicians that like overtime and don't want changes to eliminate this need. Regardless, the management team will need to be attentive of these issues in order to not let them interfere with the changes necessary to keep the plant efficient.

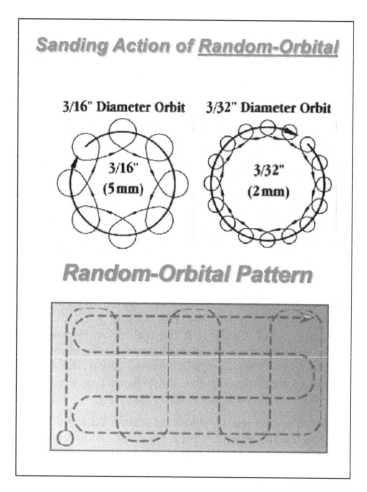

FIGURE 19.1
DA Sanding

Hire Your Replacement

Many managers have mixed emotions about this. The perception is that if you train your replacement, you will be replaced. This is definitely a possibility; however, if you follow the plan outlined in this book, you won't be losing your employment; you will more than likely be promoted or moved to another area to improve it as well.

The benefits of training your replacement are there are two people with the same agenda and skill set, which allows you to be absent without hurting the plant, and makes for a smooth transition if you retire or are moved to a different position. Another benefit is that people are all different, and having another person who has a better understanding of the plant will offer suggestions and answers that you might not have thought of previously. This will offer stronger decision capabilities.

The person you choose should be one of the top performers in your organization. Their personality is key. They must be able to work with all levels of people in the organization, and they must be loyal to you. It really is not necessary that they understand the specifics of managing the plant since you will be coaching them. This is not something that happens quickly as you both have existing positions in the company, so you should plan on spending at least 6 months, and you might want to start this training after you are well on the path with your intended goals. You should, however, let your direct reports know that you will be selecting one individual for this training once the plant is operating to the specified goals.

Training your successor will require as much of your time as it does theirs. They must first also have someone in their area who can also assist them in their function while they are in training, and they will still be responsible for the results of their area. Once they have this person in place, you can begin their training.

Start by teaching them in the main areas of the plant, one department at a time. Often the best way to learn is to sit in on meetings with the managers of each area and observe. They should be familiar with all of the reports used in each area, and the responsibilities of the managers in charge. They need to have a good understanding of how each area operates, and how to work with the manager responsible.

Forward-Thinking

Always look at your processes, people, and materials and evaluate them for future improvements. I call this forward-thinking – looking ahead long term to try to envision what changes need to be made in order to stay competitive and to maximize profit. The following are several avenues you can use to help envision possible improvements long term:

1. Trade periodicals – These magazines frequently stay on top of new discoveries in your trade.
2. Vendors – They visit many facilities, some identical to yours, some in other markets that use similar technologies.
3. Trade Shows – You can learn new processes and technologies from other managers and vendors. You will often get to work with or see new equipment at these shows.
4. Sub-committees formed from the employees in your plant. – Your employees also learn different things from other places they have worked, and can offer new ideas for improvements.

5. Equipment audits – Your equipment manufacturers or vendors can offer advice for new improvements you might not be familiar with.

6. Interns – This is a great way to get the latest knowledge at a fair price, and from temporary employees who generally work hard to make show their value. Often they are more than happy to accept a position when finished with their degree.

Conclusion

Although it is not always easy to manage, change is necessary in order to stay competitive especially concerning the advance of new technologies. Change management often requires that you acquire the ability to manage down when production levels decrease. It is also a great idea to train one of your best employees to perform your job as this will help ensure your plant operates effectively and efficiently when you are not able to manage it, and will allow for additional ideas to help you improve the facility.

Terms

Blind test
Bottleneck
DA backup pad
Dual Action sander
Forward-thinking
Gloss readings
Grain size sequence
Mechanic's Creeper
Profilometer
Strobe effect
Trade periodicals
Working in the zone

Additional Reading

www.businessinsider.com/concept-flow-help-get-in-zone-work-brad-stulberg-2017-9

www.nortonabrasives.com/sga-common/files/document/aam_cat_uk_2018_web.pdf?t=27394648

Section 6

Forward-Thinking

20

Preventing Major Issues

Introduction

There are two departments in composite manufacturing that warrant additional attention for making sure composite facilities operate with the best possible chance for success, and the greatest avenues for maintaining the lowest operational cost possible. They are: Facility Maintenance and Quality Assurance.

Having visited over 500 different composite facilities, it became apparent these departments are considered an afterthought at best for about 75–80%. Most facilities do not operate without some type of maintenance and quality program, but most do lack procedure, preventative maintenance, organization, and technical skill. Many facilities do not include a lab for testing raw materials, and do not believe they need one. This chapter highlights the impact from not placing a strong focus on these departments, how these departments should function, and what tools are necessary to assure their success.

What Does Maintenance Look Like When Things Don't Look Right?

It is always easier to point out flaws than to determine the root cause. Too often in manufacturing, especially in new businesses, the main focus is on producing product or attainment to the schedule, and since the facility may also be new, less attention is given to incorporating a maintenance schedule into the process. Maintenance becomes reactionary. Something breaks, or is damaged, and maintenance is called in a state of panic to rush through a fix, which often results in a less-than-ideal repair. The root cause is not usually addressed, and the issue surfaces again at another less-than-ideal time. Reactionary maintenance makes it incredibly difficult to allocate appropriate labor to the maintenance department, schedule their time, balance their workload, and address root causes. In large facilities, which have several plants performing diverse functions, utilizing a central maintenance team for reactionary maintenance often equates to additional downtime, bottlenecks, late deliveries, and additional management time expended. Read Example 20:1 to understand the impact of reactionary maintenance.

Example 20:1 No Time for Maintenance

A large composite facility that encompassed six different plants, including Tooling, Lamination, Mill, Assembly, Small Parts, and Upholstery, had one maintenance team to handle all issues. Maintenance at this facility was purely reactionary, and was performed based on the level of urgency. Issues with the lamination plant took precedence over most other plants at the facility since downtime often meant lost product, additional scrap, safety, and manufacturing issues that affected other plants downstream in the product process.

The upholstery plant at this facility had several machines, including a Gerber CNC to cut the upholstery fabrics, another CNC used to cut core materials for substrate, a large slitter for cutting fabric rolls into smaller widths, about 30 industrial sewing machines, and a few small machines including bandsaws and sanders. The Upholstery plant supplied product to the main Assembly plant, which completed the finished product for the customer.

The Gerber was about 1 year old, and the maintenance team did not have anyone skilled to perform maintenance on it. Consequently, no preventative maintenance had been performed when it went down and stopped cutting the necessary fabrics needed to complete the 700 finished upholstery parts required daily. It was now a crisis since production had started a few hours prior, and determining which patterns per part were not cut, locating these paper patterns, and hand-cutting them would require additional manpower and excessive overtime. Even if this was possible, the finished upholstery product would not get to the Assembly plant on time. The Gerber had to be fixed, and it had to happen within one shift to alleviate issues downstream.

Maintenance was called and was informed of the urgency. Since they were dealing with other issues in Lamination, no one was sent to the Upholstery plant for over an hour, and after they finally arrived, it was determined that the issue was beyond their capability to repair. After exploring what the plant maintenance department could do regarding the Gerber, a decision was made to call the manufacturer's repair technician. This was a cost of over $2,000 USD, and an additional day waiting on this individual to arrive.

The Upholstery Manager, not wanting to repeat this same scenario in the future, located the instruction manual for the Gerber, and determined what preventative maintenance was required to keep the

Gerber at optimum performance. Not wanting to be held hostage by the facility Maintenance department, the Upholstery plant utilized their own workers on off hours to maintain the machine. Consequently, the machine rarely had issues from this time forward.

The ideal solution would have been to have the maintenance team take responsibility of understanding what maintenance was required to prevent the Gerber from failing at less than optimal times, work with the upholstery plant to assure they used and cared for the Gerber as required, and gained the necessary knowledge to work on the machine.

This is just one of several examples of how inefficient maintenance processes can disrupt manufacturing, costing the company additional funds. Additional examples I have witnessed in the past or been exposed to at other locations I have visited are:

- Gelcoat and chopper guns that are not maintained or calibrated by either the employees or the maintenance personnel, and result in quality defects on the product including catalyst drips, porosity, and heat.

- Flotation foam integrity issues from lack of preventative maintenance on the gun, which resulted in lost product and quality issues (see Example 15:13).

- Agitation systems that failed, which resulted in the thixotropic (thix) index issues and caused 10 boats to be replaced due to sag.

- Lamination facility ceiling tiles missing or damaged, and rain water dripped onto the product resulting in laminate cure issues.

- Multiple airline leaks, which result in approximately $200 USD per leak annually in electricity increases.

- Lamination resin heaters that failed, and resin was unable to spray correctly from the equipment resulting in a 60-hour labor downtime, and loss of product.

- A 25-year replacement part that was not ordered in advance for a critical machine to manufacture product, and when it went down, it cost the company several thousands of USD in lost sales, presented opportunities for competitors to gain access to new business, required overtime when the part arrived, and consumed management time dealing with the exception.

- Grinding room dumpsters not getting handled correctly, resulting in a fire. Some may say this is not a maintenance issue, but a manufacturing issue. I believe it is both: when the company burns to the ground, everyone is out of work.

How Should a Maintenance Department Function, and What Is Required to Accomplish This?

Maintenance is generally not noticed by most until something breaks, but the goal of a successful Maintenance department is to head this off before it happens, or at least be ready with appropriate parts to minimalize downtime and product loss in the plants. As with anything we manage, we need to understand what specifically we are responsible for, and what it will take to accomplish this. The simple goal for maintenance is to maintain the safety of the facility, and protect the integrity of all that it contains. The below list highlights some of the responsibilities a composite Maintenance department will need to achieve.

- Understand and keep accurate records and information on every piece of equipment and structure at the facility, and either have personnel to maintain this equipment, or have outside sources to call.
- Work with production managers to set up preventative maintenance schedules on all equipment and facility structures as recommended by the equipment manuals or other sources.
- Work with the production managers to develop a matrix of equipment issues, in comparison to levels of urgency to repair in an emergency situation.
- Develop a process for reporting exceptions and non-compliance.
- Track spending on all equipment, and report this monthly to the area managers.
- Keep additional repair parts ready for emergency situations, and general repair. Replacement gun heads are easier to change on the floor, which allows the maintenance technician time in the maintenance repair area to perform a better job on the failed equipment.
- Hire diverse maintenance technicians, including those with either license or experience as electricians, mechanics, plumbers, and chemical handling.
- Process all handling of waste from the facility.
- Keep the maintenance repair areas clean and organized. Utilize a work order system to help track labor and material cost. Balance this to justify the employee head count in the maintenance department.
- Promote training for the maintenance and care of equipment and facility structure.
- Treat the facility and all it contains as if you own it, and have zero tolerance for abuse, misuse, theft, or vandalism.

- Work with Manufacturing, HR, and the Safety team to determine a shutdown and start-up schedule for normal operations and emergency situations.

What Does Quality Assurance Look Like When Things Don't Look Right?

Quality Assurance is not always viewed in a positive light by manufacturing. Often they are blamed for defects they do not directly produce. They are sometimes thought of as lazy, stupid, or intentionally trying to inflict pain on the workers or managers through their inspections. I have seen cases of inspectors getting their tires slashed in the parking lot, and some extreme cases where fights have ensued over a disagreement concerning an item on the inspection sheet. It isn't too hard to determine when a composite quality assurance department is not functioning as quality assurance. The following are some good examples of what this looks like:

- Excessive issues in repair resulting from the lamination processes including cracking from the pulling station, ineffective rolling, catalyst drips, and heat. This is suggestive of a void in inspection practices in the lamination areas. Quality personnel should be working with the management team to identify issues, and help find the root cause to eliminate the issue permanently.

- Frustrated repair employees and inspectors concerning a lack of standards for finish quality. This happens when there are either no quality standards, or the standards are not being adhered to. Finish quality on gelcoat surfaces experiences this more than functional items as discrepancies are frequently left up to the inspector to determine.

- Gelcoat, Foam, and Resin issues resulting from chemical noncompliance either from the chemical manufacturer, or from how the gelcoat was handled by the plant prior to use. This suggests that the quality department is not actively testing raw materials at receiving or throughout the manufacturing process.

- The absence of a tracking system to identify defects left per employee, and a process to either train, retrain, or reprimand noncompliance. This suggests that QA is not working with the plant management team and employees to help them work toward zero defects in a way that will help them accomplish this goal.

- Excessive or consistent fiberglass or gelcoat damage due to engineering defects. This suggests QA is not working with Engineering to help eliminate issues with quality.

- Continual warrantee complaints for the same issues. This suggests no communication between QA and plant management, or a lack of attention from the plant management team regarding repetitive issues.

- The absence of a Quality Control Lab or one that is missing critical measuring tools. A composite company cannot operate effectively long term without a lab. The tests performed in the lab give an insight to a skilled lamination manager, and allow them to adjust if required and allowed. These tests are also a good record for later issues that may arise through warrantee. They also help to make certain questionable loads, which can cause issues, are not delivered to the plant.

- The Quality Assurance department answers to the Plant Manager, and their inspections are influenced by other factors including production and bonus more than the quality standards. Often this is the case, and if the Plant Manager has a strong focus on quality, it may not be an issue. If they do not, the inspector will be placed in a no-win situation, and will more than likely try to appease the plant manager.

- The absence of a "checks and balance" system for upholding quality. Often the rule of "I wrote it down, you have to fix it" prevails when a system for disagreements regarding defects arise. Training the inspectors, employees, leads, supervisors, and managers on quality standards can be difficult regarding finish quality with gelcoat or other composite materials because everyone looks at finish quality with a different pair of eyes, and we all have different vision with or without a light to help us see.

- A lack of support from Senior Management regarding quality. If management is lax in their regard to quality, they can expect the employees and the QA team to eventually follow suit. I have seen this happen at even very large manufacturers of composite parts.

How Should a Quality Assurance Department Function, and What Is Required to Accomplish This?

The Quality Assurance department should strive to be excellent at helping the plant to achieve zero quality defects as defined by the company quality standards. This zero-defect grade card should ultimately come from their customers, and while it is difficult if not impossible to always achieve this especially in regard to the fit and finish of composite finish gloss and quality, it should still be the goal. This department is in a unique position to deliver bad news, and at the same time work with the production team to help determine and eliminate the root cause of the defect. Below are some points to consider when implementing an effective Quality Assurance department.

- The Quality Assurance Manager will benefit from having a strong background in the chemical areas of a composite facility as the majority of the warrantee defects originate from fit and finish, and the expensive ones to fix originate in the lamination areas.

- The inspectors will benefit from working in the manufacturing areas they inspect. Lead people are ideal candidates to hire inspectors from as they understand the process, and should already possess a high work ethic. They are familiar with management's expectations, and are able to lead a team. This is important as their new role is not just to uncover issues, but to work with the plant to help find the root cause, and eliminate the issue completely.

- A formal set of quality standards including pictures in most cases is essential if you are going to be successful at getting employees to adhere to a standard. All employees have to understand what is and is not acceptable in order to recognize a defect. If this isn't available, they will become frustrated with differences of opinion, and either quality will suffer, or cost will increase from overworking the issues.

- A Quality Assurance lab for testing raw materials is essential to head off issues before they become costly. The lab should include several pieces of equipment including the following:

Viscometer: This tool is used to check the viscosity of the chemical, which helps to determine the thixotropic (thix) index, and will determine if the resin will sag on the vertical sides during production processes. It also helps to determine how it will flow in the spray processes. See Figure 20:1 for a picture of what a Viscometer looks like.

- Gel timer: This instrument measures the amount of time a catalyzed sample reaches the gel stage. This gives the lamination department a good idea of how long they will have to roll out air bubbles before the product starts to harden.

- Burn oven: This piece of equipment measures the ratio of fiberglass to resin which helps determine for the manufacturer if the desired range was reached. If the product has too much resin, the strength will suffer; too much glass increases the risk of dry areas. The oven acts like a kiln and burns off the resin leaving the fiberglass. Weights are taken with the resin, and after the burn to determine the ratio.

- In addition to the above, QA will find it beneficial to also have a gram scale for weighing the burn sample, a temperature gun for checking peak exotherm, a barcol impressor for measuring hardness on cured product, mil gauges for checking if proper levels of raw materials have been applied, a gloss meter to determine the level of gloss on a completed part, a magnifying light for looking at defects to help determine chemical

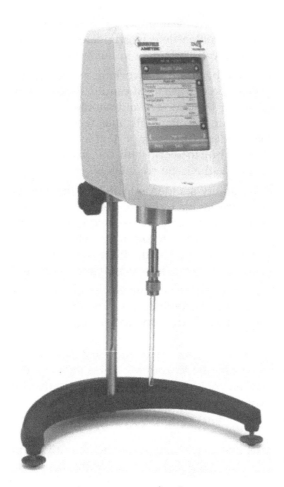

Viscometer/gel timer

FIGURE 20.1
Viscometer

integrity, and a profilometer for measuring the scratch depth and consistency of the scratches from sanding in order to determine which grain
size to use to eliminate additional work. Pictures of some of these
products can be found in Figures 20:2, 20:3, and 20:4.

- A defined process of Checks and Balances for inspection that is based
  off of the quality standards, which everyone understands and follows. This process utilizes the following format:

FIGURE 20.2
Gloss Meter

1. A supervisor calls for an inspection of a product.

2. The inspector inspects the product to assure it is manufactured to standard.

3. The employee has a disagreement with an item on the inspection, and either is or is unable to resolve it with the inspector.

4. If unable, the supervisor checks the issues, and either agrees with the inspector or the employee.

5. If the inspector does not agree, they involve the quality assurance manager.

6. If the supervisor and the QA manager do not agree, the plant manager is called.

7. If the plant manager and the QA manager do not agree, the General manager is called.

The point of this is not to create additional management time spent as issues usually never get to the level of management, but to highlight the need for everyone to be concerned with the quality standards, and to understand what the quality expectations are. This helps to keep inspectors from varying too far away from the standards, and puts the responsibility

Profilometer

FIGURE 20.3
Profilometer

Barcol Impressor

FIGURE 20.4
Barcol Impressor.

for quality on everyone's shoulders. Too often concerning gelcoat issues, the gelcoat is overworked, which does not improve finish quality.

- The QA department should keep records of all issues by product number, department, and employee name if possible. This information should be kept in a product file for warrantee purposes, and should be used to help the plant determine areas where training or corrective measures are required.

- Quality Assurance should work with engineering to correct issues for the plant that cause defective product.

Conclusion

Maintenance and Quality Control are two departments that support manufacturing processes if utilized correctly. Although often hard to attach a complete reduction figure to, they both reduce cost in manufacturing if utilized correctly. Most companies do not take advantage of these assets, and do not recognize them as a cost reduction avenue. If utilized correctly, they will reduce downtime and rework.

Terms

Barcol impressor
Bottlenecks
Burn oven
Catalyst drips
Gel timer
Gelcoat
Gloss meter
Heat
Mil gauges
Peak exotherm
Porosity
Preventative maintenance
Profilometer
QA Checks and Balances
Reactionary maintenance
Resin heaters
Root cause
Temperature gun
Thixotropic index
Tooling
Viscometer

References

www.brookfieldengineering.com/products/viscometers/laboratory-viscometers/
dve-low-cost-digital-viscometer

www.zoro.com/mahr-federal-inc-pocket-surface-gage-10micron-probe-2191800/i/
G0463760/feature-product?gclid=Cj0KCQiAnNXiBRCoARIsAJe_1cpKjgRW8
qufVPiM7i_ZLi5fXc5XDE9_tqusoEtS3kmAB_hAzF-hP_IaAkoMEALw_wcB

www.amazon.com/Horiba-3014035117-Model-IG-331-Gloss/dp/B002P8GLWQ/
ref=asc_df_B002P8GLWQ/?tag=hyprod-20&linkCode=df0&hva
did=241968535606&hvpos=1o3&hvnetw=g&hvrand=2885772533482828324&hvpo
ne=&hvptwo=&hvqmt=&hvdev=c&hvdvcmdl=&hvlocint=&hvlocphy=9012313&hv
targid=pla-457618451921&psc=1

www.abqindustrial.net/store/index.php?main_page=product_info&product
s_id=476&cvsfa=3652&cvsfe=2&cvsfhu=32343734&gclid=Cj0KCQiAkfriBRD1AR
IsAASKsQJDgf-ys2DPGwlbjpPI4cWjs9edtYjIT1fCvnmQqUT9Yk_
iUG8L__UaAhyTEALw_wcB

Glossary

5S principles: A system for cleaning and organizing work centers for efficient work.

Agitation system: A mechanical system to mix a chemical such as resin or gelcoat in order to keep the chemical within manufacturing consistency.

Air voids: A defect in composites where air bubbles that are trapped between the layers of laminates and the gelcoat during the manufacturing process leave a weaker product that can break when pressure is applied.

Audits: An evaluation of a process, area, or plant to determine issues that when corrected would reduce operational costs.

Balancing work flow: The process of assigning labor to work centers in a way that eliminates bottlenecks and creates a continuous flow between processes until the product is complete.

Barcol Impressor: A portable precision instrument which measures the hardness of a surface.

Bill of Materials: A list of materials needed to manufacture the finished product.

Blind Test: A test performed, which utilizes technicians that do not realize all of the specifics of the test.

BOM: An acronym or initialism used to represent the Bill of Materials.

Bonus program: Money shared by a company with their employees generally given for going above expectations.

Bottleneck: A congestion in the process that causes a disruption to the normal process flow.

Buffing compound: A chemical mixture which includes abrasive grain, and is used to take out fine surface scratches leaving the finish quality glossy.

Buffing pad: A pad used on a buffing tool which is used to take out fine surface scratches.

Buffing swirls: Scratches created by a buffing pad that has been either contaminated or used improperly.

Build book: Detailed instructions for how to manufacture a product.

Build schedules: A list of composite materials with instructions, specifically for the lamination process, used to manufacture a composite part.

Burn Oven: Equipment used to burn resin from a fiberglass plug for determining a resin to fiberglass ratio.

Burn Test: A test consisting of a cured fiberglass cutout, which is weighed then heated in a kiln or burn oven, to remove all resin leaving the fiberglass alone, and weighed again to determine a resin to glass ratio.

Calibrate: A process for checking the accuracy of the spray equipment in order to be certain it is providing the exact characteristics it is set to give.

CAPEX: A term, which stands for Capital Expenditure, and represents company funds used to improve or replace physical assets such as equipment or building upgrades.

Capital: Company funds

Catalyst drips: A defect experienced in laminates when catalyst leaks from the gun onto the product during the manufacturing process.

Chain of command: The hierarchy of a management structure, which allows for employees to follow or answer to their immediate leader.

Change orders: A change to the original work order for manufacturing a specific product, usually initiated by a customer.

Character skills: Personality characteristics an individual has, which can be good or bad, and that identify their ability to be trusted to work ethically and independently.

Charge code: A code given to a specific department within a company, that is used to identify cost and location for products purchased from a vendor or in-house manufactured parts.

Chemical integrity: The ability of the chemical to retain the specific composition detailed on the lab report provided by the chemical company when delivered to the plant.

Chop strand: A small bale of fiberglass roving generally used with equipment to produce the first layer of laminate on top of the gelcoat or coating.

Chopper gun: Equipment used to cut fiberglass roving into smaller lengths, and spray it with resin and catalyst onto the mold to produce a fiberglass reinforced plastic (FRP) part.

Chopper operators: Technicians who operate the chopper gun.

Closed Press Molding: The laminates are placed dry into the mold, the two part mold is clamped together, and the chemicals are injected into the cavity.

CNC: Computer Numerical Control, which means a computer is used to send instructions to the cutting machine.

Cockpit floors: An area of a boat that is partially open and lower than the top of the deck.

COGS: Cost of Goods Sold. This includes all costs to manufacture and produce a product that is sold.

Composite: Any solid product that is made from different parts or elements. Good examples are concrete, the fiberglass reinforced plastic part of a boat or fiberglass bathtub.

Construction Manuals: Books that detail how a product is intended to be manufactured.

Consultants: Industry experts who offer their advice for a fee.

Continuous Flow: During the manufacturing process, products or parts are moved individually and continuously from one station to the next until complete.

Core the decks: The process of adding structural materials such as foam or wood during the lamination processes to give support to the deck of a fiberglass boat.

Cost Analysis: Determining the total cost of a process or product change as a comparison to understand if a change is financially beneficial.

Crew Chief: A position in manufacturing that describes a person responsible for leading a team of employees to perform the daily tasks. Often called Team Leaders or Crew Leaders.

Crew Leaders: A position in manufacturing that describes a person responsible for leading a team of employees to perform the daily tasks. Often called Crew Chiefs or Team Leaders.

Cure: The point at which a plastic composite product is considered a solid material.

Cutting room: The area of a composite manufacturer where fabrics are cut to a pattern.

DA Backup pad: A support pad that is secured to tool, and holds an abrasive intended to sand a product.

DA scratches: Scratches produced by the abrasive used on a Dual Action Sanding tool. These take the shape similar to a fishhook.

Data points: Points on a graph that dictate specific measurements of an activity such as production efficiencies.

Deck: The top half of a fiberglass boat.

Deck Core Kit: A patterned kit cut from support materials that is designed to fit a specific deck, and is generally shipped with a placement drawing for ease of manufacturing.

Delamination: A failure in composite manufacturing that occurs when the laminated layers of materials separate.

Disc Contamination: A defective condition on an abrasive sandpaper, which results from not cleaning the sanding discs periodically during use, and the swarf or buildup forms on the disc creating larger, undesirable scratch sizes.

Discipline Motivation: One of the three types of motivation (Positive, Negative, & Discipline) that is used to initiate positive change. It is the least favored of the three; is generally only used for extreme cases involving safety, security, or insubordination; or when the other two have failed.

Dual Action (DA) sander: A tool used with an abrasive to sand or refine a product, and utilizes a sanding pattern that involves two motions: a full circular, and at the same time, a small orbit within the circle.

Durometer: An instrument for measuring hardness.

Electronic Data Interchange (EDI):   The process for sending and receiving data, such as a purchase order from a manufacturer to a vendor electronically.

Equal Employment Opportunity Commission (EEOC) Charge: An accusation of discrimination or harassment made by an employee about believed treatment they have received from their employer.

Employee Retention: The ability to keep the employees that you hire.

Employee Turnaround (turnover):   A figure that represents how often the company has to replace an employee in comparison to how many employees the company retains on average in a given time period.

Engineering changes: Changes to the manufactured product documented by the Engineering department.

Exercise in futility: Engaging in an activity that will never produce a beneficial outcome.

Filament Winding: A process which applies fiberglass strand wet or dry onto a rotary mold or mandrel.

Fillers: Products used in compounds to add lubricant and other benefits for the abrasive grain.

Finishers: Employees who work on the cosmetic aspect of the gelcoat to increase the quality of the finish.

Fit & Finish: The overall aesthetics of a product.

Fixed resources: A resource such as land that does not change as production increases.

Fixtures: Items used to hold two laminated parts in place for accuracy during the bonding process.

Flanges: The area of a laminated part that is trimmed as waste.

Flotation Foam: A two-part chemical when combined creates a solid foam that fills a cavity and is a requirement for flotation in certain boats under 26 feet in length.

Forward-thinking: The act of understanding what the end result will be for an issue or opportunity long term. The ability to see things long term.

Fiberglass Reinforced Plastic (FRP):   A composite made from fiberglass material, a resin, and a catalyst to cure.

File Transfer Protocol (FTP):   Communication between a computer and a server to transfer files.

Finished Goods: Completed product that has not yet left the facility or work center.

Gantt chart: A chart that is utilized to illustrate a sequence of events.

Gauge components: The instruments on a dash or helm panel.

Gel Timer: An instrument used to test how long a chemical such as resin or gelcoat takes to get to the gel stage when removing air bubbles in the laminates will be difficult.

Gel-coaters: An employee who sprays gelcoat.

Gelcoat: Pigmented resin that acts as the base coat and gives color to the finished product.

Gelcoat Repair: The process of correcting an issue in the gelcoat.

Geometric nonskid sheets: A part of the plug that provides a patterned surface called Nonskid for slip resistance in the finished part once pulled from the mold made from the plug.

Gloss Meter: An instrument that measures the gloss of a product.

Gloss Readings: The data taken from using a gloss meter on the surface of a product to determine the gloss quality.

Grain sequence: A systematic grain size progression to achieve a desired surface quality when using abrasives.

Grain shed: The process of grains loosening from the backing of an abrasive during the sanding process.

Grinders: In the composite industry, a grinder is an employee who works in the Trim & Grind areas, and removes the flanges and sharp edges as well as cuts all access holes.

Grinding booth: The area of a composite facility where all grinding on the composites is performed.

Hardtop assemblies: The assembly that generally covers the helm area or console of a boat, and includes the component parts.

Hatch: A lockable cover to an opening on the top side of a deck.

Heat: A quality issue that occurs when excessive temperatures during cure adversely affect the gelcoat giving it a distorted look from shrinkage or imprint of the fiberglass material.

High-Performance Boat: A powerboat that is manufactured for its speed on the water.

Hull Schedule: The start schedule for a Lamination plant.

Hulls: The lower half of a boat.

Human capacity: The capability of a person to give much more in work than expected.

Human Resources: The department in a facility that handles recruiting, labor issues, benefits, etc.

Inboard/Outboard (I/O) Motors: The motor of a boat that resides inside the boat, and is attached to a drive unit that resides outside the hull.

In-house customers: The departments or plants in a facility that receive manufactured or produced products from another department or plant in the same facility.

In-mold coating: The application of a liquid coating to a molded part while it is still in the mold.

Inefficiencies: Activity or measures that are not acceptable.

Infusion: A composite process that applies resin into the materials utilizing a vacuum bagged liner.

Initiator: A chemical which reacts with the resin to achieve cure.

Injection: A closed mold process used in high-speed composite operations. Materials are injected into the closed mold.

Internal customers: Same as in-house customers.

Information System (IS) department: The department in a corporation that processes information in an effort to keep other areas functioning efficiently.

ISO Auditor: A person, often certified, who is responsible for checking to be certain that all stated documentation for how things should be done, actually are being done as stated.

Jigs: A tool that holds products together and provides a guide for another tool necessary to produce an accurate part.

Just-in-time (JIT): Manufacturing that allows product to be manufactured individually as needed.

Kanban: A manufacturing process that provides parts used in the production processes as needed for Just-In-Time (JIT) manufacturing.

Kiln: An oven that reaches temperatures in excess of 1500, and is typically used for pottery.

Labor-intensive industry: An industry that predominantly relies on people to perform the majority of the work instead of machinery or robots.

Lamination: The area of a composite factory that processes layers of chemicals, reinforcement, and core to manufacture the FRP product.

Laminators: Employees who work in the lamination area, and manufacture the FRP product.

Learned helplessness: A condition that people exhibit when frustrated and don't see any way to improve their situation.

Legacy systems: An outdated computer system.

Lost-time-accidents: Accidents occurring at work that require time away from work.

Machine-able Syntactic putties: Sprayable putties used in manufacturing plugs for molds.

Macro number: Often the top-level number that does not give details for the combined numbers.

Maintenance Repair & Operations (MRO): A term used to describe the products used to support manufacturing, but don't usually become part of the completed product.

Markers: The name given for the data a CNC uses to cut a set of pre-planned patterns.

Maslow's Hierarchy of Needs: A theory Abraham Maslow gave for how individuals are able to be motivated.

Material Savings bonus: A bonus given for using less shop supplies than allowed to complete a product.

Material variance: The difference between how much material is allowed to produce product, and how much was actually used.

Milestone: The achievement of a difficult goal or exceptional event.

Mil gauges: A tool used to measure the applied thickness of wet material such as gelcoat or resin.

Mill Shop: The area of a facility that processes and manufactures the cabinetry, and core products used in the other plants at a facility.

Mold Availability: Often molds are shared between a few models of product, which presents an additional challenge for scheduling the lamination start.

Molds: A very thick laminated or metal tool used in the lamination process to manufacture a composite part.

Monthly Incentive: A bonus given on a monthly basis for performance.

Morale: The level of satisfaction and agreement in respect to work atmosphere and relationships.

Negative motivation: One of the three main types of motivation, which involves taking something away to motivate an individual, but not as a punishment.

Non-disclosure agreement: An agreement between two parties to protect confidential information in an effort to keep the intellectual assets of a company protected.

Non-financial Rewards: Rewards given that have little or no monetary value to the receiving individual.

Open Mold: A composite process, which utilizes manufacturing on a mold that is not covered during the process.

Operational Costs: All of the costs associated to owning a business.

Organizational Chart: A chart that depicts the hierarchy of management in a corporation or business.

Organizational skills: The ability to structure workload and business to attain the most ideal outcome in the least amount of time.

Overhead: Costs a business incurs that are hard to include in the cost of goods sold such as electricity, shop supplies, and maintenance.

Overtime: Labor spent in excess of the normally scheduled workday and usually at a premium cost.

Peak exotherm: The point at which a laminate generates the most heat before starting to cool.

Performance bonus: An incentive or bonus money paid due to positive performance achievement.

Performance efficiencies: The measurement of how productive a department is in comparison to the measurement of how productive they should be.

Performance standards: The amount of labor assigned to perform a specific task as stated by the engineer department.

Personal Protective Equipment (PPE): Equipment meant to protect an employee such as safety glasses, gloves, paper suits, respirators, steel toed shoes, and hard hats.

Plugs: The inverse of a mold, which is used to manufacture a mold.

Point-of-Use Cabinets: Storage cabinets positioned closest to the area where the products they hold are needed in manufacturing.

Porosity: Tiny trapped air bubbles in gelcoat generally resulting from spray issues.

Positive motivation: One of the three motivational techniques that rely heavily on supporting acceptable behavior in an effort to repeat or improve the behavior.

Post-cure: Resin that continues to cure after the product has been pulled from the mold resulting in poor finish quality.

Preventative maintenance: Maintenance that has been scheduled in advance of equipment failure.

Process Flow: The path that a product takes from beginning to completion.

Procurement: The department in a facility that is responsible for attaining the necessary purchased products needed for manufacturing.

Production rate: The amount of product produced in a specified time period.

Production schedule: A detailed schedule that maps out key dates for completion of product as it moves through each station or area.

Profilometer: A precision tool which measures the roughness of a surface.

Profit margin: A ratio attained by dividing the net income from products manufactured by the net sales of those products.

Project Software: Computer software used to map out the details of a specific plan.

Promoter package: Addition chemicals that are added to resin or gelcoat to change its properties.

Pulling station: The area of a composite facility that is responsible for removing the product from the mold.

QA Checks and Balances: A system in Quality Assurance that helps to train all employees on the quality standards the company has predetermined.

Quality standards: The quality expectations a company has for the product it produces.

Reactionary maintenance: Unscheduled maintenance that takes place after something breaks.

Regression analysis: An in-depth analysis of data, which combines several variables to give a more accurate picture of how they influence a dependent variable.

Resin Heaters: Heaters that keep the resin in the desired temperature range for manufacturing when the ambient temperature is below that range.

Resin pumps: Equipment that pumps the resin and catalyst at set levels for control of manufacturing.

Resources: Assets of a company that can be used to keep the business operating.

Responsibility matrix: A diagram showing the responsibility of leaders and how they relate to other leaders.

Return on Investment (ROI): The measurement of profitability of an investment.

Roller doors: Large roll-up doors in a facility.

Root cause: The initial reason a problem exists.

Resin Transfer Molding (RTM): A closed mold process that includes a pre-formed dry fiber insert and pumped resin between two clamped molds.

Service Corps of Retired Executives (SCORE): A non-profit organization that matches mentors with businesses to help them improve their business and reduce operational costs.

Seamstress: An employee whose position is to sew.

Seat base assemblies: Fiberglass components that are used in the cockpit of a boat to hold a seat.

Shop Floor: Production or Manufacturing floor space where the operations are performed.

Shrink test: A test to show the percentage of shrinkage of a chemical after cure.

Sign off: A Quality Control check that assures a process has passed inspection.

Skillsets: The ability to perform a task as required.

SPEC: Specification, as outlined by the engineering team.

Spin molding: A laminating process utilizing a rotary mold that is filled with an uncured composite, and is spun for a specified amount of time till cured.

Spray additives: Chemicals added to the gelcoat.

Spur: To clean the buffing pad, or a tool that cleans a buffing pad.

Squash molding: Clamping two molds with parts and a bonding agent in them together to make one part when pulled.

Staging: Organizing work product and supplies according to work center.

Stand-alone documentation: Documentation that is easily understood without the need for additional documentation.

Standard Labor Hours: The cost to the company for one employee to give one hour of work. This includes benefits, tax, and overhead.

Stockroom: The area of a facility where the purchased product used in manufacturing is received and often stored.

Stress cracks: Cracks in laminated product that result from stress to the part.

Stringer system: The ribbed core part of a boat hull that gives it strength and stability.

Swarf: Dust from the product being sanded.

Tack: In an upholster shop, this represents the action of stapling the materials to a substrate.

Temperature gun: Also called a heat gun, it is an instrument that can determine the temperature of products that reach temperatures up to 400 F. It is generally used with products that are in the curing process.

Thixotropic Index: A reading taken on resin to determine if it will be able to hold its shape.

Throughput: The amount of product a manufacturer is able to produce with given resources in a specific time period.

Time study: A documented study of the amount of time it takes for a specific task to be completed.

Tooling: A term that is often used to describe the area of a composite facility, which repairs and maintains the molds. This term is also often used to describe the molds in a composite manufacturing facility as well.

Trade shows: Industry-specific events where manufacturers, vendors, and other associated folks share ideas, new products, and services.

Trend-line graphs: A graph which depicts the rate at which something is increasing or decreasing over a specified time period.

Tuition reimbursement: A benefit companies use which pays for college expenses and is generally relevant to the industry they are part of.

Vacuum-assisted device: A tool that utilizes a vacuum system for removing dust as it is sanding or grinding.

Variance: A measurement of how far something is away from its intended goal.

Vendor managed inventory: Using vendors or distributors to account for, procure, and stock necessary supplies for manufacturing.

Viscometer: Equipment that measures the viscosity or thickness of resins.

Work in Process (WIP): Products that have not yet been used to complete the finished product.

Work Flow: The process that products move through a manufacturing plant until complete in the finished product.

Work instruction: Directions for manufacturing a specific product.

Workload: The amount of work a department or station is able to complete in a given time.

Workman Compensation Insurance: Insurance that covers medical expenses and some wages for employees who are hurt while working for an employer.

Workstations: A defined area for specific tasks to be performed in a plant.

References

http://everestexpedition.co.uk/everest/seven-attributes-to-climb-everest/

www.investopedia.com/terms/c/capital.asp

http://leansixsigmadefinition.com/glossary/5s/

www.rocklin.k12.ca.us/staff/aparker/WEB/Sayings%20of%20Poor%20Richard.pdf

https://hbr.org/2012/09/how-to-respond-to-negativity.html

www.simplypsychology.org/maslow.html

www.verywellmind.com/what-is-learned-helplessness-2795326

www.merriam-webster.com/dictionary/composite

www.eeoc.com

https://resources.workable.com/tutorial/calculate-employee-turnover-rate

www.google.com/search?safe=active&client=safari&channel=iphone_b
m&ei=ntBhXNDJD9Ht5gKJ4IrwDg&q=filiment+winding&oq=filiment+wind
ing&gs_l=psy-ab.3.0i10l10.102878.105194.105410...0.0.0.96.1288.16......0....1.
gwswiz.......0i71j35i39j0i67j0i131i67j0i131j0.dYF2ufAz1EQ

www.nmma.org/lib/docs/nmma/cert/exams/Basic_and_Level_Flotation.pdf

https://searchnetworking.techtarget.com/definition/File-Transfer-Protocol-FTP

www.compositesworld.com/zones/lfrt-injection-molding

www.myaccountingcourse.com/financial-ratios/profit-margin-ratio

https://coventivecomposites.com/explainers/resin-transfer-moulding-rtm/

www.score.org/frequently-asked-questions-about-score

https://compositeenvisions.com/raytek-mini-temp-thermal-heat-gun-
thermometer-1508.html

http://neutron.physics.ucsb.edu/docs/optical%20epoxy/viscosity_thixotropic_in
dex.pdf

Index

5S principles, 39, 177

A

ACMA, 110–111
Agitation system, 62, 167, 177
AIM Supply, 116
Air voids, 153–154, 177
American Bath Group, 109–110
Audits, 94–95, 106, 121–124,
    162, 177

B

Balancing work flow, 132, 177
Barcol Impressor, 171,174, 177
Bill of Materials, 4, 12, 15–17, 24, 26, 28,
    93, 130, 132, 177
Blind Test, 159, 177
Bonus program, 147–150, 177
Bottleneck, 4, 23, 158, 166, 177
Buffing swirls, 104, 177
Build Books, 15, 24, 148, 177
Build schedules, 95, 177
Burn oven, 171,177
Burn Test, 29, 171, 177

C

Calibrate, 6, 57, 167, 178
CAPEX, 56, 95, 178
Catalyst drips, 167, 169, 178
Chain of command, 75, 80–81, 178
Change orders, 140, 143, 178
Character skills, 61–65, 68, 178
Charge code, 113, 178
Chemical integrity, 150, 178
Chop strand, 27, 178
Chopper gun, 56, 153, 167, 178
Chopper operators, 19, 153, 178
CNC, 78, 92, 114, 117, 119–120, 166,
    178, 182
Cockpit floors, 28, 178

COGS, 93, 131, 178
Components, 5, 71, 91, 98, 140, 185
Composite Research, Inc, 119, 122–123
Construction manuals, 24, 178
Continuous flow, 92, 177, 179
Core the decks, 28–29, 179
Cost analysis, 42, 179
Crew Chief, 131, 179
Crew leaders, 29, 31, 131, 179
Cure, 108, 115, 167, 171, 179–181, 184, 185
Cutting room, 40, 53, 117,179

D

DA scratches, 104, 179
Deck, 28–29, 153–154, 178–179, 181
Deck Core Kit, 28–29, 179
Delamination, 29, 179
Disc contamination, 54, 104–105, 179
Discipline motivation, 70–72, 79, 179
Dual Action sander, 159, 166, 179
Dull spots, 104, 105
Durometer, 114, 179

E

Employee Retention, 13–15, 28,
    143, 180
Engineering changes, 100, 132, 180

F

Fillers, 105, 180
Finished goods, 84, 91, 95, 97
Finishers, 19
Fixtures, 4, 15, 71, 93, 180
Flanges, 6, 140, 180
Flotation foam, 119, 167, 180
Forward-thinking, 68, 157, 161, 180
FRP, 24, 41, 63, 97–98, 180

G

Gauge components, 140, 180

Gel Timer, 171, 180
Gel-coaters, 19, 180
Gelcoat, 6, 28, 41, 53–54, 71, 84, 104–111,
        122–123, 140, 158, 167, 169–170,
        174, 180
Geometric nonskid sheets, 114, 181
Gloss meter, 171, 173, 181
Grain sequence, 104, 106, 159, 181
Grain shed, 104, 181
Grinding booth, 64, 69, 81

H

Heat, 167, 169, 181
High-Performance Boat, 115, 153, 181
Hull Schedule, 142, 181
Human capacity, 9, 57, 181
Human Resources, 20, 61, 76, 91, 93, 181

I

In-mold coating, 110, 181
Initiator, 110, 181
Internal customers, 97, 181

J

Jigs, 4, 15, 71, 93, 182
Just-in-time, 4, 100, 117, 182

K

Kiln, 104, 171

L

Labor charts, 20, 132
Lamination, 5, 6, 27–29, 46, 54, 63–65,
        71–72, 84, 86, 94, 105–106, 121,
        142–143, 153–4, 166–167,
        169–171, 182

M

Machine-able Syntactic putties, 114, 182
Managing down, 157
Markers, 117, 182
Material Savings bonus, 150–152, 182
Material variance, 17, 24, 45–46, 84, 182

Measuring tools, 129, 170
Mil gauges, 171, 175, 182
Mill Shop, 5, 97–98, 182
Molds, 4–6, 71, 95, 114–115, 140, 183
Monthly incentive, 84, 183
MRO, 110–112, 183

N

Nano Boats, 113
Negative motivation, 67, 69–71, 79, 183
Norton, 53, 104–106, 123

O

Open mold, 6, 67, 183

P

Peak exotherm, 171, 183
Performance Efficiencies, 12, 15–16,
        84, 183
Performance standards, 45, 154, 183
Plugs, 114–115, 183
Point of Use cabinets, 111, 183
Porosity, 110, 167, 183
Positive motivation, 68–69, 72, 78, 184
Post Cure, 107–109, 184
Personal protective equipment (PPE), 19,
        91, 184
Preventative maintenance, 4, 62–63, 94,
        117–119, 165–168, 184
Process flow, 5, 143, 184
Procurement, 4, 18, 91, 93–4, 184
Profilometer, 159, 172, 174, 184
Project Software, 45, 48, 184
Pulling station, 27, 169, 184

Q

QA Checks and Balances, 172, 184
Quality standards, 4, 104, 158,
        169–173, 184

R

Reactionary maintenance, 63,
        165–166, 184
Regression analysis, 137, 184

Resin heaters, 167, 184
Resin pumps, 63

S

Safety, 3, 12, 19–20, 30, 51–52,
        64, 76, 84–85, 91, 103, 152,
        157–158, 166, 168–169
SCORE, 121, 185, 187
Shrink test, 114, 185
Spin molding, 6, 185
Staging, 37, 185
Standard Labor Hours,
        16, 185
Stress cracks, 28–29, 185
Stringer, 5, 104, 185

T

Thixotropic index, 167, 171, 186
Time study, 111–112, 186

Tooling, 3, 5–6, 113–15, 166, 186
Tuition reimbursement, 152, 186

V

Vacuum assisted device, 53, 186
Variance 17, 24, 26, 45–46, 84, 182, 186
Vallen 111–113
Vendor managed inventory 111, 116, 186
Viscometer 171, 172, 186

W

WIP 140, 142–143, 186
Work flow 41, 132, 177, 186
Work instruction 86, 186
Working in the zone 157
Workload 20, 46, 57, 157, 165, 183, 186
Workman Compensation Insurance
        83, 186
Workstation 41, 186

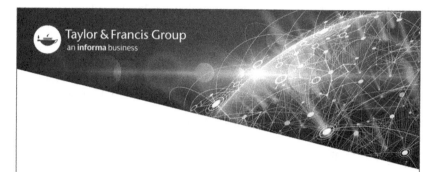